FURTHER DEVELOPMENTS IN TURBULENCE MANAGEMENT

FLUID MECHANICS AND ITS APPLICATIONS
Volume 19

Series Editor: R. MOREAU
MADYLAM
Ecole Nationale Supérieure d'Hydraulique de Grenoble
Boîte Postale 95
38402 Saint Martin d'Hères Cedex, France

Aims and Scope of the Series

The purpose of this series is to focus on subjects in which fluid mechanics plays a fundamental role.

As well as the more traditional applications of aeronautics, hydraulics, heat and mass transfer etc., books will be published dealing with topics which are currently in a state of rapid development, such as turbulence, suspensions and multiphase fluids, super and hypersonic flows and numerical modelling techniques.

It is a widely held view that it is the interdisciplinary subjects that will receive intense scientific attention, bringing them to the forefront of technological advancement. Fluids have the ability to transport matter and its properties as well as transmit force, therefore fluid mechanics is a subject that is particulary open to cross fertilisation with other sciences and disciplines of engineering. The subject of fluid mechanics will be highly relevant in domains such as chemical, metallurgical, biological and ecological engineering. This series is particularly open to such new multidisciplinary domains.

The median level of presentation is the first year graduate student. Some texts are monographs defining the current state of a field; others are accessible to final year undergraduates; but essentially the emphasis is on readability and clarity.

For a list of related mechanics titles, see final pages.

Further Developments in Turbulence Management

edited by

K. KRISHNA PRASAD
Eindhoven University of Technology
The Netherlands

Reprinted from Applied Scientific Research, Volume 50, Nos. 3/4, 1993

SPRINGER SCIENCE+BUSINESS MEDIA, B.V.

Library of Congress Cataloging-in-Publication Data

Further developments in turbulence management / edited by K. Krishna Prasad.
p. cm. -- (Fluid mechanics and its applications ; v. 19)
ISBN 978-94-010-4745-6 ISBN 978-94-011-1701-2 (eBook)
DOI 10.1007/978-94-011-1701-2
1. Turbulence--Congresses. 2. Drag (Aerodynamics)--Congresses.
I. Prasad, K. Krishna. II. Series.
TA357.5.T87F88 1993
620.1'064--dc20 93-19297

ISBN 978-94-010-4745-6

Printed on acid-free paper

TABLE OF CONTENTS

Preface

This special issue of *Applied Scientific Research* returns to the theme of turbulence manipulation with a view to reduce drag. It provides a selection of papers presented at the Sixth European Drag Reduction Meeting held at the Eindhoven University of Technology at Eindhoven in November 1991. The procedures adopted in the preparation of this issue was similar to the ones used in the two earlier proceedings volumes brought out by Kluwer Academic Publishers in 1990 and 1991.

A significant departure of the present issue from the earlier proceedings volumes is the virtual absence of papers on LEBU's indicating their unfulfilled promise of drag reduction. However, there is one paper that considers their potential for manipulating heat transfer from surfaces. All the rest of the papers are devoted to surface-initiated manipulators. The bulk of the present issue is concerned with riblets. An attitude shared among the authors is that riblets reduce drag. The main question addressed thus is the mechanism responsible for this behaviour of turbulent flows. The papers reflect the general thrust of modern day turbulence research in their concern on coherent structures on the one hand and modelling on the other hand. The latter approach combined with the detailed analysis of the viscous sublayer provided in one of the papers can be the way forward for the design of optimized riblet shapes to yield increased drag reduction. The d-type roughness/cavity has continued to attract the attention of the research community as evidenced by the presence of two papers on the subject and a third one looking af the effectiveness of dilute polymer and surfactant solutions flowing over such surfaces.

The Sixth European Drag Reduction meeting was hosted by the Eindhoven University of Technology by providing all the infrastructure necessary for any such meeting. The European Research Community on Flow Turbulence and Combustion (ERCOFTAC) provided a few scholarships to enable young researchers to participate in the meeting; Kluwer Academic Publishers provided a grant that came in handy to meet some of the external costs of the meeting. Lastly, Addy Schwarz van Manen looked after scores of details in connection with the meeting's organization. To all these I owe a special thanks.

Eindhoven, February 1993

K. Krishna Prasad
Guest Editor

Applied Scientific Research **50**: 189–213, 1993.
K. Krishna Prasad (ed.), *Further Developments in Turbulence Management.*

Bursting and structure of the turbulence in an internal flow manipulated by riblets

S. TARDU[1], T.V. TRUONG & B. TANGUAY*

*Swiss Federal Institute of Technology-Lausanne; 1015-Lausanne, Switzerland (*author for correspondence)*

[1]*Institut de Mécanique de Grenoble-Laboratoire des Ecoulements Géophysiques et Industriels; B.P. 53-X, F-38041 Grenoble-Cédex, France*

Abstract. The buffer layer of an internal flow manipulated by riblets is investigated. The distributions of the ejection and bursting frequency from the beginning to the middle part of the buffer layer, together with high moments of the fluctuating streamwise velocity, u', and its time derivative are reported. The profiles of the ejection and bursting frequency are determined and compared using three single point detection schemes. The effect of the riblets on the bursting mechanism is found confined in a localized region in the buffer layer. The multiple ejection bursts are more affected than the single ejection bursts. The skewness and flatness factors of the u' signal are larger in the manipulated layer than in the standard boundary layer. That, also holds true for the flatness factor of the time derivative, but the Taylor and Liepman scales are not affected. The spectrum of the u' signal is altered at the beginning part of the viscous sublayer.

Nomenclature

u_τ = Friction velocity
ν = Viscosity
l_ν; f_ν = wall scales ν/u_τ; u_τ^2/ν
y = Vertical distance to the wall
Δz = Spanwise extent
$(^+)$ = Variable normalized with wall scales
u = Velocity; u' = Turbulence intensity
h, s = Height and width of the riblets
f_e = Ejection frequency
f_b = Bursting frequency
f_{BME} = Frequency of the Bursts with Multiple Ejection
f_{BSE} = Frequency of Single Ejection Bursts
$S_{u'}$ and $S_{du'/dt}$ = Skewness factor of u' and its time derivative
$F_{u'}$ and $F_{du'/dt}$ = Flatness factor of u' and its time derivative

Abbreviations:

SBL = Standard (non-manipulated) Boundary Layer
MBL = Manipulated Boundary Layer
BME = Bursts with Multiple Ejections
BSE = Bursts with Single Ejections
VITA = Variable Interval Time Averaging technique
u'-l = u'-level technique
mu = Modified u'-level technique

1. Introduction

There is a general consensus between several research groups on the efficiency of the riblets in reducing drag reduction in internal or external flows with or without adverse pressure gradients (Choi, 1989, 1990; Truong and Pulvin, 1989, Rohr et al., 1989) at high subsonic and transonic (Walsh and Anders, 1989; Coustols 1989) and supersonic (Gaudet, 1989) speeds (see Savill, 1990 for a recent review). However, it is quite interesting to note that there is a considerable discrepency between the results aimed at determining the possible effects of riblets on the structure of the near wall turbulence (Wallace and Balint, 1988). This is particularly true when the previously published results on the eventual reaction of the bursting process in the internally manipulated boundary layer are compared: although a number of very detailed investigations have been conducted, including single and multipoint turbulence quantities, it is striking to observe that the existing data ranges from an increase of the bursting frequency by a factor of 8 (Choi, 1989) to a decrease by about 30% (Gallagher and Thomas, 1984). The first explanation of this considerable disparity is of course due to the differences in the detection techniques and detection parameters applied, as well as the size and location of the probes used. Another point seems also to be important to point out: the maximum drag reduction obtained by the riblets is about 8% (Savill, 1990); this implies that the corresponding increase of the internal time scale (i.e. ν/u_τ^2) is also only 8%. If we assume that the unique effect of the riblets is probably the decrease of the bursting frequency – i.e. not the modification of the 3D coherent structures or the shear stress distribution within the burst (as suggested by Pulles et al., 1990) or other possible effects, and if the bursting frequency still scales with inner variables (ν and u_τ^2) the detected decrease should not be larger than 8% which is within the scatter of the previously published results concerning the bursting frequency in the standard boundary layer (SBL hereafter). Consequently more objective detection techniques, if possible, are needed to identify the modification of the bursting mechanism. Another difficulty which arises comes from the fact that the effect of the riblets on the flow structure is confined in the buffer and viscous sublayer (as clearly shown by Choi, 1989), and this implies careful and detailed measurements near the wall.

VITA is the most commonly used detection technique, certainly because it is a well known and documented scheme in the SBL and requires only a single signal. Walsh (1982) found no significant alteration of the bursting frequency nor of the contribution to the overall turbulent intensity during and between bursts. They conclude that the main effect of the riblets is to modify the wall streaks distribution in space and reduce the burst intensity. Hooshmand et al. (1983) reported conflicting results when the integration time T_v in the VITA scheme is chosen relative to the local u_τ computed from the sublayer profile or their modified Clauser method. Their data in the SBL case however, is well below the data of Blackwelder and Haritonidis (1983) indicating that the spanwise extend of their probe is too large. Gallagher and Thomas (1984) used a sufficiently small probe ($\Delta z^+ = 9$) and reported that the number of bursts on the riblet surface at $y^+ = 15$ was 30% lower than that detected on the flat surface. The same amount of decrease is also reported by Savill (1987) based on flow visualizations. The most interesting result reported by Gallagher and Thomas (1984), is the increase in size of the spanwise structures by about 15–30% obtained by long time spanwise correlation distributions at zero time delay. This is contrary to the smoke-wire visualisations of Hooshmand et al. (1983) which reported no apparent difference in the average spacing between structures. Their dye visualizations indicated also that the spanwise spread of the dye is constrained within the rib walls and the interaction of the flow between adjacent grooves is weak.

The increase of the spanwise streak spacing was also clearly shown by Bacher and Smith (1985). They identified the streak spacing by visual counting and they reported a 40% increase. They did not observe a modification of the bursting frequency, but they noted a wiggle preceding and following the sharp acceleration of the conditional patterns. They attributed these oscillations to the stabilization of the streak behaviour below $y^+ = 15$.

Choi (1986, 1987, 1989, 1990) reported very detailed measurements concerning the structure of the turbulence in the internally manipulated layer both with and without adverse pressure gradients. He detected the bursts using the instantaneous wall shear stress signal by using VITA with a threshold equal to one. The spanwise extend of the probe was about 10 wall units which is comparable with Gallagher and Thomas, 1984. He reported an increase of the bursting frequency by a factor of èight, but a decrease of the r.m.s. intensity of the turbulent wall shear stress by the same amount. The time scale of the conditional VITA averages at the wall was found to decrease by a factor of two so that the intermittency increased by a factor of four. The hot film was flush mounted within the spanwise distance of the riblets, and how the heat transfer to the substrate affected the measurements and the detection of the structures remains an open question. An increase in size of the near wall coherent structures was once more observed. A systematic increase of the skewness and flatness factors of the streamwise fluctuating velocity was also reported.

Another conflicting result came from the research group of the Tech. Un. of Eindhoven (Pulles et al. 1989). When VITA was applied to the streamwise $u'(t)$ signal at $y^+ = 14$, the frequency of the detected events was found to increase in the MBL (the increase is about 20% for a threshold $k = 0.4$). But when applied to $v'(t)$ signal, VITA gave a decrease of about the same amount. The physical meaning of these results is open to discussion. Their conclusion was that the single point measurements are inadequate to detect changes of the bursting process in the internally manipulated boundary layer. This conducted the same research group to investigate the bursting using quadrant techniques by uv measurements (Schwarz-van Manen et al. 1990). They identified individual ejections and grouped them into bursts. Their closest measurements point to the wall is $y^+ = 10$. They report that the bursting frequency in the SBL decreases with y^+ which is in contradiction with Tiederman (1988) and Bogard (1982) who reported constant bursting frequency for $y^+ > 15$. Although the scatter of their data is important they reported a 30% decrease of bursting frequency.

An important question concerns the effect of the riblets geometry on the coherent structures. To the authors knowledge, the only study which dealt with different groove types is the one conducted by Pulles et al. (1989) who used 8 different geometries. They did not report a spectacular effect of the groove geometry on the ejection frequency. On the other hand, some of the riblets which were used so far have particular geometries (for example Choi, 1988) and it is difficult to conclude from the published data a common trend because of the differences in the detection techniques and locations, the sizes of the probes, etc.

The main conclusion of this short review is that not only are the measurements on the behaviour of the fine structure of the turbulence in MBL contradictory, but they are also generally limited to the outer part of the buffer layer and detailed analysis in the inner part of the buffer layer ($y^+ < 15$) are needed in order to elucidate the effect of the riblets, since there is a general concensus that (contrarily to LEBUs) the main effect of the riblets is confined near the wall. On the other hand it is known that in canonical turbulent boundary layer, the bursts originate from the low-speed streaks at $y^+ < 12$ (this is called the development stage of the bursting process in this study) and it is asked here if the process that generates the break-up of the low-speed streaks in the buffer layer is influenced by the presence of the riblets or not.

The first aim of this study is to use several detection and grouping schemes to identify bursting structure in the wall layer part of an internally manipulated layer. Several other characteristics such as the power spectra, high moments of the streamwise fluctuating signal together with statistics concerning the time derivatives, small and intermediate scales (Taylor and Lieppman scales) are also investigated.

2. Experimental set-up

The experiments were conducted in the closed loop wind tunnel with low turbulence level (0.5%) described in details by Pulvin (1989). The working section is a two dimensional channel, 2 m. long with two fixed transparent side walls and 2 aluminium flat plates with hydrodynamically smooth (or riblets) surface and adjustable in angle. The whole set of data was taken here in a zero pressure gradient. The measurements were taken at $x = \delta$ upstream of the end of the riblet plate.

Symmetrical V type Hoecht riblets were used in this study (Fig. 1). The height and width of the riblets were $h = s = 0.01'' = 0.25\,\text{mm}$.

Most of the part of this study was investigated with a mean centerline velocity of $U_c = 15\,\text{m/s}$, corresponding to a Reynolds number of $Re_\delta = 27500$. The drag reduction was evaluated using the momentum balance in the same way as Pulvin (1989) and found in the present working conditions ($h^+ = s^+ = 9$) to be $\Delta C_f = 0.94$. This value corresponds quite well to the previous measurements of Pulvin (1989). The

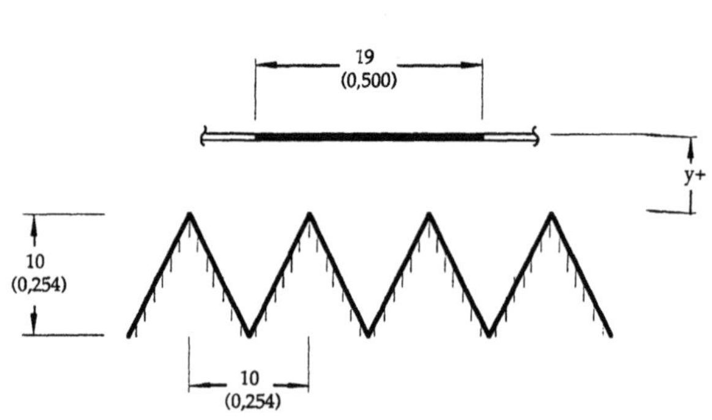

Fig. 1. Cross section of the riblets together with the hot film probe. The dimensions are in wall units. The values in brackets are in mm.

top of the grooved plate is used as the origin. The u_τ measured here would correspond to the average value from the peak and valley values. Therefore the shift above of the velocity profiles (which is slight here) is less pronounced than that of Hooshmand et al. (1983). This means that, if we use the notion of the protrusion height, i.e. the virtual origin that the outer flow sees, as defined by Bechert et al. (1986), the y^+ computed should be shifted towards the centerline by an amount of $0.1-0.2\,h^+ = 0.9-1.8$ (this value depends on the geometry of the riblets and no verification was made in this work; see the viscous analysis due to Bechert et al., 1986). This would of course accentuate the effects presented in this study. On the other hand, since the spanwise extent of the probe is 2 times the riblets spacing, the spanwise location of the probe should not change the detected characteristics, i.e. the probe is large enough to have average characteristics corresponding to the buffer layer, and can not detect the more detailed configuration of the flow within the ribs. (The case studied here is similar in this aspect to the previous measurements of Fulachier et al., 1987). The use of a smaller probe would clarify the dynamical process as a function of its lateral position with respect to the groove pattern. There is a real gap concerning this last point in the literature that future research should carry on.

A subminiature TSI 1260-10 hot-film probe is used. The spanwise length of the probe is $\Delta z^+ = 19$ in wall units; although it is somewhat large, it may be assumed that it is in the acceptable limit to avoid the spanwise averaging of the detected structures (Blackwelder and Haritonidis, 1983).

The sampling frequency is set equal to $f_v = u_\tau^2/v$ (20 kHz for $U_c = 15$ m/s). The signal was low pass filtered by a Krohn Hite filter at $f_v/2$. The data was transferred to the host computer and analyzed in several ways.

The total duration of each record was $T_r = 4600\,\delta/U_\infty$ and this is enough to ensure the convergence of statistics up to 4-th order moments (Klewicki and Falco, 1990) and of the ejection and bursting frequency for the non-manipulated boundary layer. This, of course does not imply a priori that the statistical convergence is ensured in the manipulated boundary layer, since the development in high order moments for different surfaces can be quite different. To this end, the convergence of the results was checked especially for the flatness factor of the time derivatives which is the most likely affected by convergence problems. Such a test is shown in Fig. 2 for the Standard Boundary Layer (SBL) and the Manipulated Boundary Layer (MBL) at $y^+ = 7.3$. Only the results concerning the 4-th order moments are shown, since the lower moments converge more rapidly. Fig. 2a and 2b show respectively the flatness factor of the fluctuating streamwise velocity, and of its time derivative versus T_{data}/T_r where T_{data} is the duration of the truncated record. It is seen that the results change by less than 1% once $T_{data}/T_r > 0.6$. The results concerning the flatness factor $F_{du'/dt}$ of the time derivative du'/dt converge at $T_{data}/T_r > 0.8$, therefore, less rapidly as it would be expected (Fig. 2b). Similar results are obtained for other sets of data investigated in this study. Note finally that the record length is almost 5 times longer than the value $T_r = 1000\,\delta/U_\infty$ recommended for the convergence of bursting frequency (Blackwelder and Haritonidis, 1983).

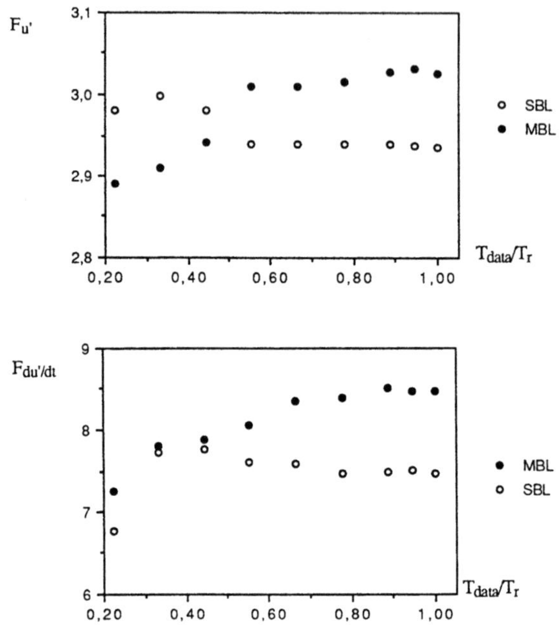

Fig. 2. Test of convergence of high order moments; (a) Flatness of the fluctuating streamwise velocity u'; (b) Flatness of the time derivative of u'. $T_r = 4600\,\delta/U_\infty$.

3. Detection and identification of the ejections and bursts

3.1. Detection of the ejections

Three independent detection techniques have been used to detect ejections, namely u'-level (u'-l), modified u'-level (*mu*), and the well established VITA (Variable Interval Time Averaging) technique. All of these detection schemes require a single signal, they essentially detect ejections but they may be slightly modified to identify sweeps as well (Bogard, 1982). Roughly speaking, the level crossing methods detect the leading edge, while VITA identifies the trailing edge of an ejection (Bogard and Tiederman, 1987). However it has to be pointed out that VITA is much more associated with near wall shear layers rather than ejections.

The rather new definition of the bursting mechanism is adopted here. Several authors indicated that the break-up of the low speed streaks may contain one or several individual lift up of the low momentum fluid from near the wall (Offen and Kline, 1975; Willmarth and Sharma, 1984; Bogard and Tiederman, 1986). They identified the break-up of a single streak as being the bursting phenomena, implying that a burst may contain one or several ejections. Recent direct simulation data also provides evidence of the existence of ejections coming close to each other (Kim and Spallart, 1987, Kline and Robinson, 1990, Robinson et al., 1988). Each one-point scheme identifies the ejections. Specific methods are therefore needed in order to group ejections into the bursts.

The detector function $D(t)$ of the m-u' and u'-l schemes is defined as follows:

$$m\text{-}u'\text{: } D(t) = 1 \quad \text{if} \quad -L_1\sqrt{\overline{u'u'}} > u'(t) > -L_2\sqrt{\overline{u'u'}},$$

$$u'\text{-l: } D(t) = 1 \quad \text{same as } m\text{-}u' \text{ but } L_1 = L_2.$$

The detector function for VITA is (Blackwelder & Kaplan, 1976):

$$D(t) = 1 \quad \text{if} \quad \sigma_v > k\overline{u'u'} \quad \text{and} \quad du'/dt > 0$$

$$D(t) = 0 \quad \text{otherwise} \quad (\,)_v = 1/T_v \int_{t-T_v/2}^{t+T_v/2} (\,)\,dt',$$

where $\sigma_v = (u'^2)_v - (u'_v)^2$.

During the whole study, the same detection parameters have been used and compared in situ with the standard boundary layer. According to Bogard and Tiederman (1986), Luchik and Tiederman (1987) and to Tardu and Binder (1989, 1991a; see also Tardu, 1988) these were later set as follows:

u'-level: $L_1 = -1.2$, $L_2 = -0.9$

mu'-level: $L_1 = -1$, $L_2 = -0.25$

VITA: $k = 0.35$, $T_v^+ = 14$.

3.2. *Grouping of the ejections*

Once the ejections are detected, they have to be grouped together, in order to identify the bursting frequency corresponding to the break-up of a single low-speed streak. Several independent grouping techniques are used in this study. Although it is a 'technical' problem, it is a crucial one, since a comparative study of the behaviour of the bursting process can clarify the effect of the riblets by those results common to several methods. Once the grouping is done, we can identify and study separately the bursts with multiple (BME), and single ejections (BSE). One of the reasons for this distinction is motivated by some recent findings that these two categories of bursts may result from different mechanisms (Tardu, 1988; Tardu and Binder, 1989, 1991a; Antonia et al., 1990).

Two different grouping methods are used here. The first one is based on the cumulative probability distribution of time between ejections (Barlow and Johnston, 1985; Bogard and Coughran, 1987). In the case of u'-l ejections the probability distribution is based on the center to center time, and in the case of m-u events, on the time between the trailing edge of one ejection and the leading edge of the next ejection

(Tiederman, 1988). These distributions have a break which marks the separation between the BME's and BSE's. A second method, originally developed by Tardu (1988) is also used and compared with the first one. It is based essentially on a specific pattern recognition technique. In an extensive study of simultaneous flow visualizations and probe measurements, Bogard and Tiederman (1986, 1987) showed that the conditional pattern of the trailing edge of the first ejection of a BME, is significantly different to that of the last ejection. In particular the maxima of the previous ejections of a BME are smaller by several orders of magnitude than the maximum of the last ejection. A typical time series of such events which would be detected by VITA is shown in Fig. 3. The method is based on a doubly iterative procedure which detects such typical events and will not be discussed here. The details may be found in Tardu (1988, 1991) and Tardu and Binder (1991a, 1992). Let us only point out that, this pattern recognition technique can be applied both to VITA, *u*-level and *mu*-level methods without modification as well as to the sweeps. It gives results comparable to the first method, and is judged to be more objective, because of the possible subjectivity made in determining the break point by use of the cumulative probability distribution.

4. Results

4.1. Profiles of the mean streamwise velocity

The streamwise velocity is scaled with the corresponding friction velocity versus the vertical distance from the wall, and the vertical distance from the top of the riblets surface in Fig. 4. It is seen that the data corresponding to the smooth plate is well represented by $u^+ = 14.5 \tanh(y^+/14.5)$ near the wall. The velocity profiles corresponding to the manipulated boundary layer is shifted slightly up in agreement with Choi (1988, 1990). An estimate of the virtual origin is done by making a least square

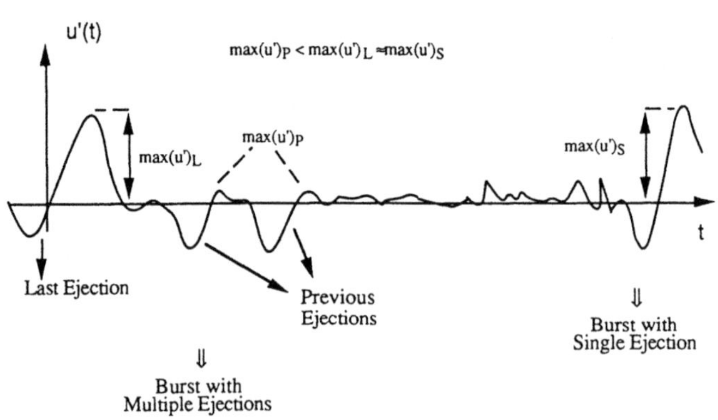

Fig. 3. Sketch of a typical burst with multiple ejections and burst with single ejection detected by the grouping method.

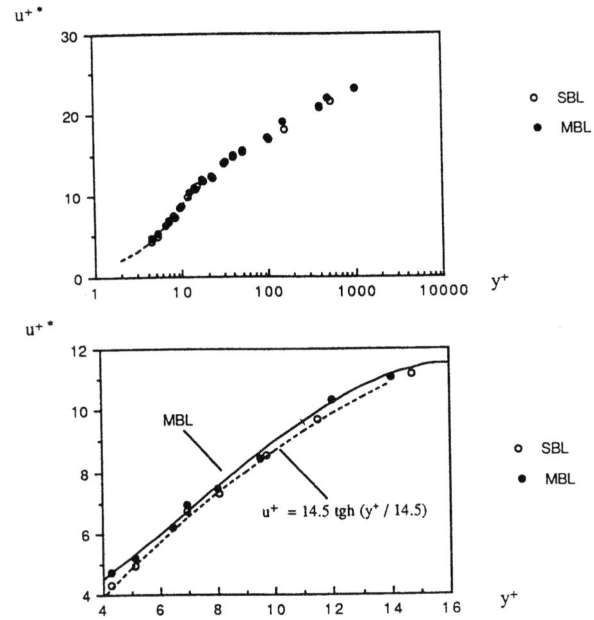

Fig. 4. Profiles of the mean streamwise velocity scaled with the corresponding friction velocity.

analysis for the first four points corresponding to the locations nearest to the riblet surface and was found $y_v^+ = 0.8$ i.e. $y_v/h = 0.1$ a value which is smaller than the one given by Bechert et al. (1986). The first reason to explain this discrepancy is that some differences may result from the spanwise distribution of the velocity profiles. The second reason comes from the small number of points taken near the wall, although the thickness of the viscous sublayer seems to increase in agreement with Choi (1988, 1990).

4.2. Profiles of the turbulence intensity

Figure 5 shows the distribution of the rms of the streamwise velocity fluctuations related to the friction velocity of the standard boundary layer. It is seen that the riblets decrease the streamwise turbulent fluctuations by an amount of 5–8% in the beginning part of the buffer layer, i.e. at $y^+ < 15$. This is a general consensus on the effect of the riblets and is in agreement with most of the previous riblets studies. Even when the rms velocity is related to the friction velocity of each case (i.e. $u_{\tau(\mathrm{SBL})}$, or $u_{\tau(\mathrm{MBL})}$) this decrease is found persistent at $4 < y^+ < 7$, a region where the bursting mechanism is found to be mostly affected.

4.3. Profiles of the ejection and bursting frequency

4.3.1. Distribution of the u-level ejection frequency

Figure 6 shows the profiles of the *u*-level ejection frequency $f^+_{e(u-1)}$ as a function of y^+.

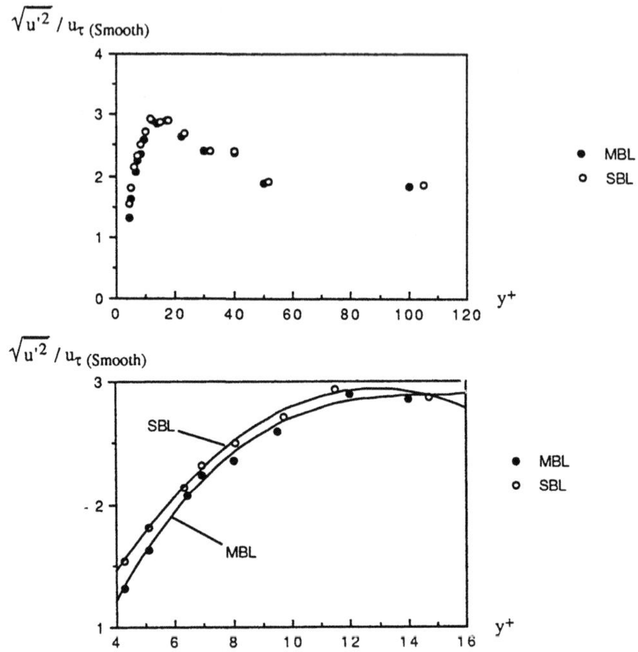

Fig. 5. Profiles of the streamwise rms fluctuations related to the friction velocity of the smooth plate.

Both f_e and the bursting frequency are non-dimensionalized with respect to the friction velocity corresponding to the SBL. It is seen, that the results for the smooth plate presented here agree in an excellent manner with the measurements of Bogard and Coughran (1987) which have used the same detection technique. This good correspondence seems to increase the degree of confidence that we may have in the results presented in this paper. Indeed, the probe used by Bogard and Coughran (1987) has a spanwise length of 10 wall units, therefore two times smaller than the one used in this study. Since the same distribution of the time-mean ejection frequency is found in both studies, it may be suggested that the spanwise extent of the most energetic eddies must be greater than $20\,\nu/u_\tau$, otherwise large discrepancies should be

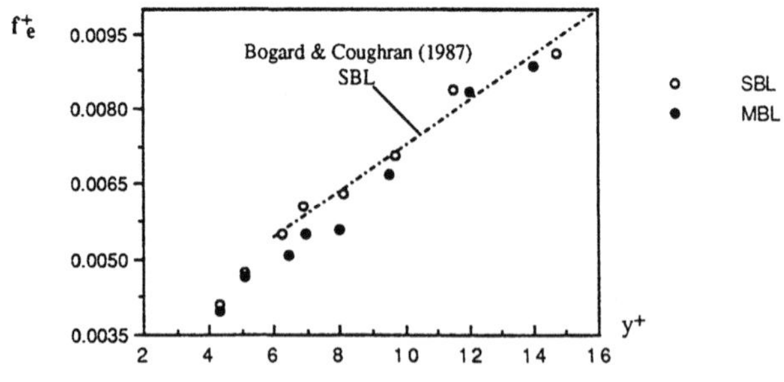

Fig. 6. Distributions of the ejection frequency versus y^+; u-level ejections.

observed because of the consequence of averaging the turbulence structures. This last point is also in agreement with Blackwelder and Haritonidis (1983), who used VITA scheme with $k = 1$ and $T_v^+ = 10$ at $y^+ = 15$. They have obtained a plateau region for the bursting frequency for $\Delta z^+ < 20$. They have related these results to the fact that the bursting mechanism is related to the break-up of the low-speed streaks which have a spanwise extent which continuously varies between $20\,\nu/u_\tau$ and $200\,\nu/u_\tau$. Thus, although the length scales of the coherent quasi-streamwise vortices which are believed to be closely related to the ejections, decrease with decreasing y^+ (the diameter of these vortices varies between 2 and $0.41y^+$; Robinson, 1990), they are found quite intermittent in the streamwise direction, and always in the immediate vicinity of the low speed streaks and the "spanwise extent of the ejection regions is (only) roughly dependent on distance from the wall" (Robinson, 1991; p. 257). Based on these arguments and comparing the results with previously published data, it may be concluded that the length of the sensor is just small enough to avoid the averaging problem in the non-manipulated boundary layer, because the length of the probe is within the distribution of the spanwise extent of wall streaks at the wall. It is however not a priori true in the manipulated boundary layer, since the effect on the spatial extent of the streaks is not clearly established up to now. However, the common belief is that the riblets increase the streak spacing. The averaging problem seems, therefore, to be less critical in the manipulated boundary layer.

To check this point, a few measurements have been carried out by using both a DISA-R11 and TSI 1260-10 probe, at $y^+ = 7$ in the SBL and MBL with different Reynolds numbers. The riblet height and spacing were between $h^+ = s^+ = 6 - 9$. The Reynolds number could not be changed beyond a certain limit in order to keep the riblets geometry sufficiently effective for drag reduction. Fig. 7 shows the ejection frequency obtained by u'-l in the MBL versus the dimensionless sensor length. The ejection frequency decreases by 27% at $\Delta z^+ = 35$, and varies by less than 7% for $15 < \Delta z^+ < 22$.

Figure 6 shows that the ejection frequency increases with y^+ because of the multiple break-up of a given streak as discussed in detail by Bogard (1982). The ejection frequency detected over the riblet surface is systematically smaller in the beginning part of the buffer layer with a maximum decrease of about 12% at $y^+ = 8$ (Fig. 8). The

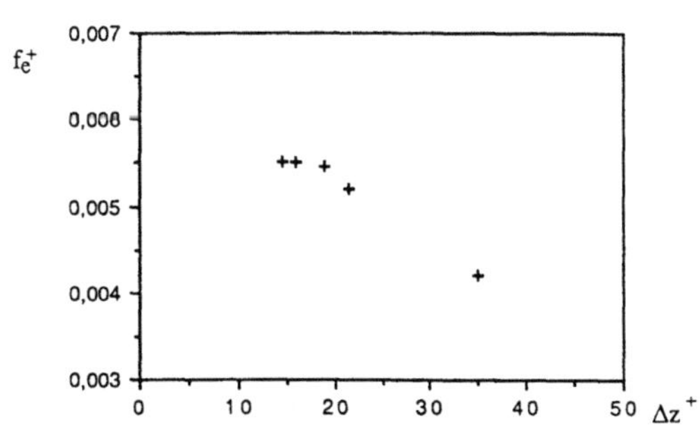

Fig. 7. Influence of the spanwise length of the probe on the ejection frequency.

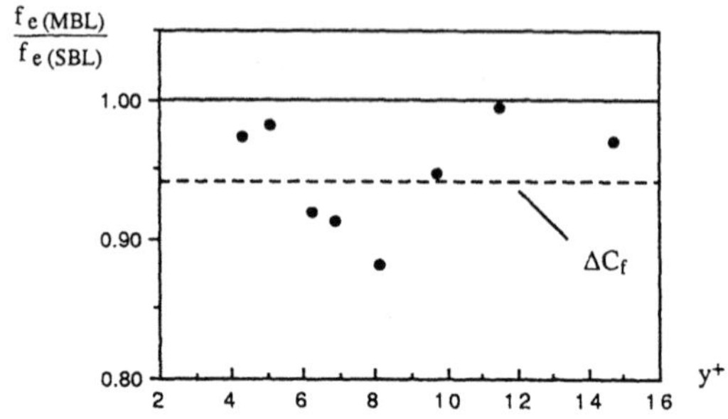

Fig. 8. Ratio of the time mean ejection frequency in MBL and SBL (u-level ejections).

difference between the ejection frequency in SBL and MBL decreases towards the outer layer. This behaviour is in some way contrary to what was observed in LEBUs, for which the principal effect was the decrease of the ejection frequency for $y^+ > 15$ (Tardu and Binder, 1991). The confined effect on the ejection frequency noted here is in agreement with the general consensus that the riblets essentially affect the inner layer with a pronounced effect in the buffer layer. The maximum decrease which takes place at $y^+ = 8$ is a rather new observation and is difficult to explain. While the ratio $f_{e(MBL)}/f_{e(SBL)}$ is systematically smaller than the drag reduction ΔC_d at $y^+ < 6$, closer to the wall the detected difference becomes negligible. This seems to be in contradiction with Choi (1988) who identified the bursts at the wall and found an increase of the bursting frequency. It is of course not a priori legitimate to extend the measurements here, to $y^+ = 0$. Another open question of course, is the validity of the detection techniques near the wall.

More physical insight on the ejection mechanism may be obtained by investigating the probability distribution of the ejection interarrival times. Fig. 9 shows such a distribution obtained at $y^+ = 8$ where the most pronounced effect on $f^+_{e(u'-1)}$ is observed. The normalized ejection interarrival time t_e^* is $t_e^* = (t_e - \overline{t_e})/\sigma_{te}$ where σ_{te} is the rms value. The rms values of the inter-arrival times are respectively $0.9\overline{t_e}$ and $1.\overline{t_e}$ for the manipulated and standard boundarylayer. The skewness and flatness of t_e are modified in the MBL, although slightly because the distributions are quite similar

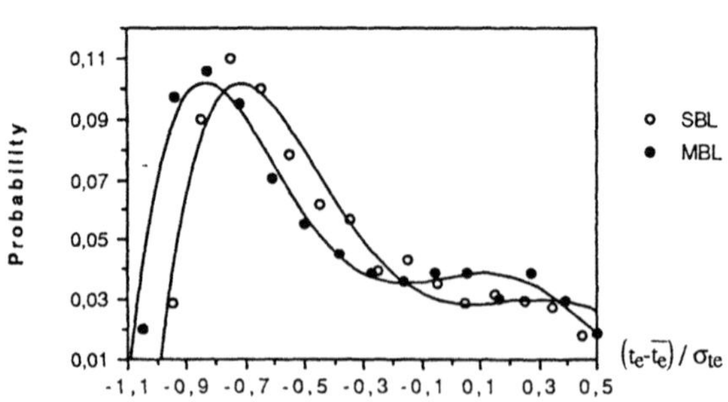

Fig. 9. Probability distribution of the ejection interrarival times; $y^+ = 8$.

(Fig. 9). Skewness and flatness of t_e over the riblets and at $y^+ = 8$ are 1.2 and 1.9 respectively compared with 0.8 and 1.1 over the smooth surface. Although there is a trend indicating that the riblets produce a more pronounced "periodic" ejection frequency, because the inter-arrival times are more intermittent (the flatness factor is two times higher in MBL), it is difficult to conclude solely from the t_e distribution whether the riblets produce a more intermittent flow or not, since this distribution does not give information on the signal between the events which may behave differently in MBL. For instance, quiescent periods of the wall shear stress signal were observed by Choi (1989) due to the partial laminarization of the viscous sublayer near the bottom of the riblets.

The distribution of the time durations of the u-level ejections was also investigated, but the results are not reported here, since no significant differences were noted. Slightly longer ejections persist in the beginning part of the buffer layer in MBL, and they are slightly shorter in the viscous sublayer (Tardu 1991).

4.3.2. Distribution of the (u′-level) bursting frequency

The grouping of the ejections is done using the second method based on the individual patterns of the detected ejections. The analysis based on the break point of the cumulative probability distribution revealed very comparable results and will not be discussed here. Let us just point out that the statistics concerning the interarrival times of the ejections up to second moments were found very similar in MBL and SBL and slight differences in higher moments were observed as it has been discussed in the previous section.

The effect of the size of the probe on the number of detected events was found less critical once the grouping is done. For instance, in a case where a larger probe ($\Delta z^+ = 37$) was used and where a decrease of 20% of the ejection frequency was observed, the bursting frequency differed only by 10%.

Figure 10 elucidates the grouping technique. The maximas of the previous ejections of a multiple ejection burst are first of all systematically negative, and are significantly different than the maxima of the last ejection or the maxima of the single ejections. There is no apparent modification of this behaviour in the MBL.

The reaction of the bursts with multiple ejections (BME), and single ejection (BSE),

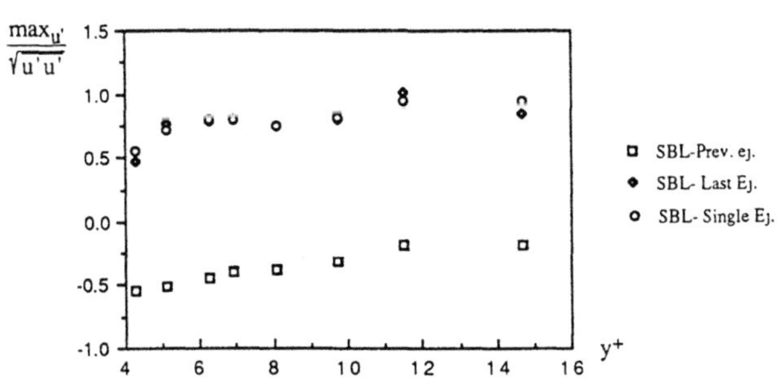

Fig. 10. Maximas of the ejections belonging to different parts of the bursts (*u*-level technique).

are studied separately in this investigation. Previous investigations (although in a different context) have suggested that these two groups of bursts may result from different mechanisms (Tardu and Binder; 1989, 1991a). In their study of unsteady wall flows, these authors have shown that, the BME's and the BSE's respond quite differently to imposed oscillations. Recently, Antonia et al. (1990) have shown that reasonable correspondence exists between the end of a burst (defined as the last ejection of a BME, or a BSE) and δ scale u-discontinuities, making a possible link between the inner and outer layer. The aim, here was to determine if these two categories of bursts behaved differently or not in MBL.

Figures 11a and 11b shows the distributions of the time mean frequencies of the BME's and BSE's in manipulated and standard boundary layer respectively. The first impression is that there are no spectacular differences when the MBL is compared with SBL. However slight but systematical alterations exist. In order to elucidate this point the ratios of the f_{BME} and f_{BSE} are plotted in Fig. 11c. First of all $f_{\mathrm{BME\,(MBL)}}/f_{\mathrm{BME\,(SBL)}}$ is smaller than one in a large part of the buffer layer including $5 < y^+ < 11$. Furthermore, this ratio is smaller than ΔC_d in most of the part of the beginning part of the buffer layer. The frequency of the single ejection bursts is found to be smaller than its homologue in the SBL, only at the closest detection station. The recovery with the corresponding SBL value is surprisingly rapid to within $\Delta y^+ = 10$. The reduction of the frequency of BME (although small, with a maximum of 15%) is larger than the reduction of the frequency of BSE.

The distribution of the bursting frequency $f_b^+ = f_{\mathrm{BME}}^+ + f_{\mathrm{BSE}}^+$ is shown in Fig. 11d. Contrary to the behaviour of the BME's and BSE's for which some stronger influence of the riblets is noted, the decrease of the bursting frequency is only of the same amount as the drag reduction. That would imply that the bursting scales with inner variables in the buffer layer. Note once more that the decrease of the bursting frequency is limited to the beginning part of the buffer layer.

The analysis of the conditional patterns dealt with several parts of the burst structure, namely the previous ejections of a multiple ejection burst, the last ejection of a BME, and the single ejection bursts. This allowed a more efficient study locked at a particular phase of the process. No significant differences were observed except at $y^+ = 7 - 8$ where a slight phase shift, and a different pattern are noted before and after the arrival of the previous ejections, and at $y^+ = 5$ where larger maxima of the last ejections are noted (Tardu 1991). This might indicate, although with some caution, that the bursts in the manipulated boundary layer are more coherent in their early stage.

4.3.3. Shear-layer (VITA and mu) events

VITA detects the near wall shear layers rather than the ejections and there is no simple relationship between these structures and the ejection mechanism. The results and comparisons presented in this section should therefore be taken into consideration with some caution.

Figure 12a shows that the shear layer events detected by VITA are less modified by

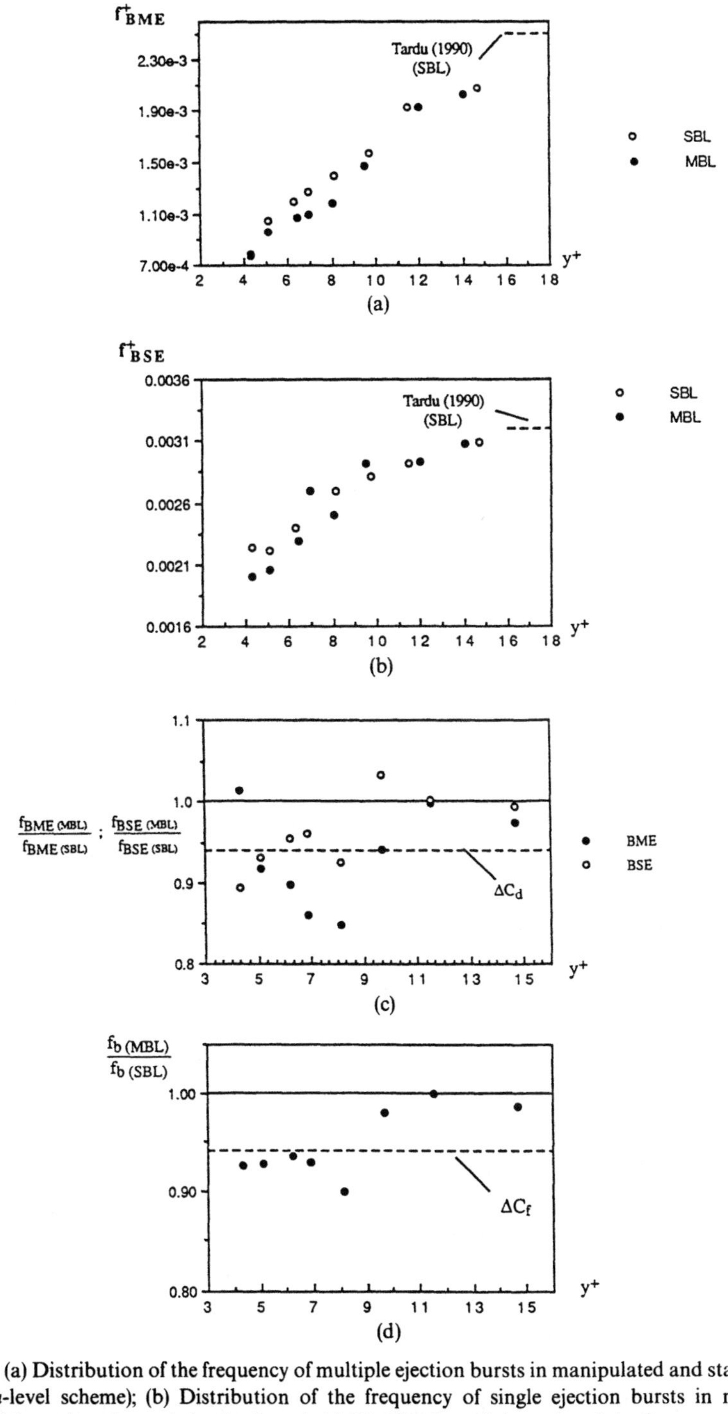

Fig. 11. (a) Distribution of the frequency of multiple ejection bursts in manipulated and standard boundary layer. (u-level scheme); (b) Distribution of the frequency of single ejection bursts in manipulated and standard boundary layer. (u-level scheme); (c) Distribution of the ratio of the frequency of multiple and single ejection bursts in manipulated to the frequency in standard boundary layer. (u-level scheme); (d) Distribution of the ratio of the bursting frequency. (u-level scheme).

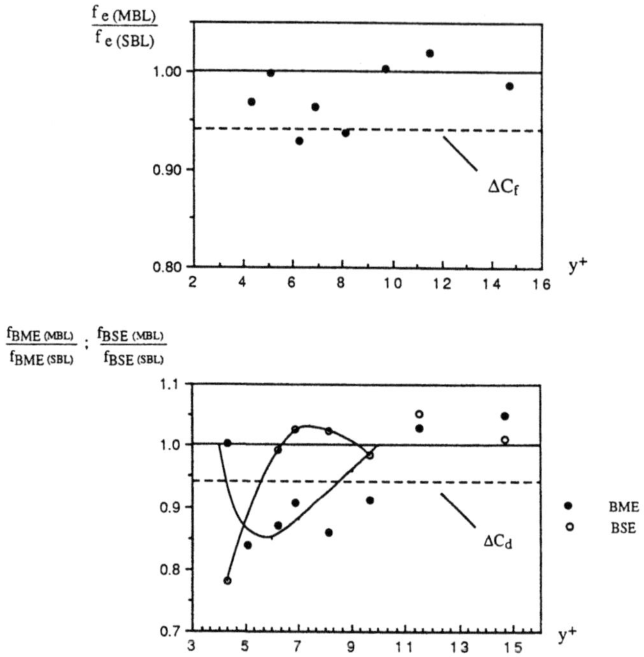

Fig. 12. (a) Ratio of the time mean frequency of the shear layer events (VITA) in MBL and SBL; (b) Ratio of the frequency of single and multiple shear layer events detected by VITA.

the presence of the riblets than the *u*-level ejections. In the manipulated boundary layer the frequency of the shear layer events is almost the same as in SBL. The difference lies within the experimental error. That is in some way similar to the LEBU's where smaller effects were observed compared to the *u*-level events (Bogard and Coughran, 1987). It is also not surprising that the modified *u*-level technique gives comparable results, since it is a level-crossing version of VITA (Luchik and Tiederman, 1987).

The maxima corresponding to the conditional patterns obtained once the grouping of the VITA and *m-u* events is done are found to be fairly similar in MBL and SBL (Tardu, 1991). Despite the fact that, the frequency of the individual VITA shear layer events, is less affected than the frequency of *u*-level ejections, the grouped shear layer events show a close similarity with the *u*-level bursts (Fig. 12b). Therefore, the same discussion done in the previous paragraph is valid here, namely that the multiple shear layer events are more sensitive to the internal manipulation than the single shear layer events. Since the multiple events correspond to the successive break-up of a single streak, that behaviour may be explained by a relative stability of the streaks so that, fewer multiple break-ups occur in the manipulated boundary layer.

Similar results are obtained by *mu*-level, as it would be expected (Tardu, 1991). As a conclusion, and despite the fact that, the three detection schemes investigated here are independent and identify different parts of the bursting mechanism, some common behaviors are noted, namely:

(i) Riblets affect the bursting mechanism at an early stage. Since here, the effect is confined to the beginning part of the buffer layer, it implies that the riblets affect the bursting mechanism in its development stage. It would therefore be interesting to apply a spatio-temporal detection technique to investigate the bursting mechanism in the internally manipulated boundary layer.

(ii) The effect is slight (10–15%), localized, and confined in the beginning part of the buffer layer. The reaction of the multiple events are different compared with the reaction of the single events.

(iii) By alterating the frequency of Multiple Ejection Bursts, the riblets probably affect also the stability of the low-speed streaks.

The sweeps could not be identified directly in this study, since this would require the $u'v'$ time series. A similar analysis was however conducted for shear layer events associated with sweeps supposing that they can be identified as decelerating VITA events. The distribution of the frequency of decelerating VITA events is very similar to the distribution of $f_{e(u'-1)}$ (Fig. 13). The effect on the multiple sweeps is clear from Fig. 14 and is similar in several aspects to the results discussed previously. A reduction of

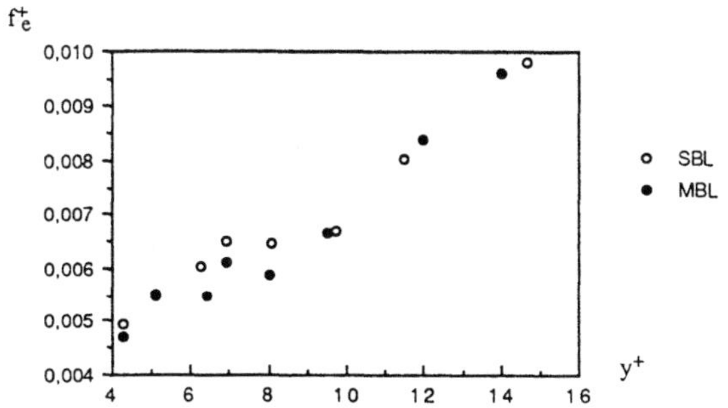

Fig. 13. Frequency of shear layer events associated with sweeps (VITA with deceleration).

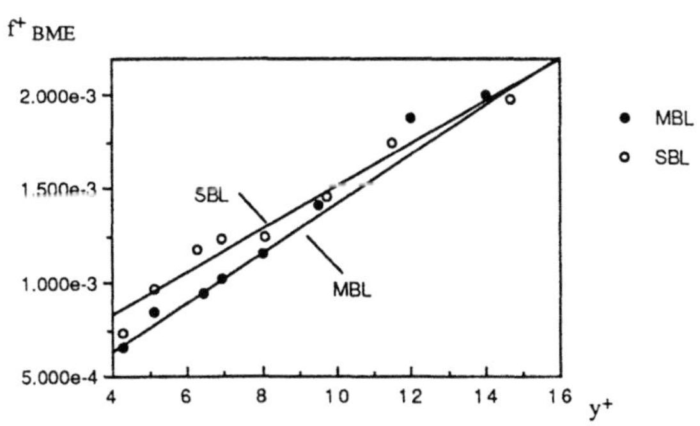

Fig. 14. Frequency of multiple sweep events in MBL and SBL.

10–20% of the frequency of the multiple sweep events is obtained between $5 < y^+ < 10$. This effect is not surprising since the sweeps are related to the ejections.

5. Characteristics of higher moments of streamwise fluctuating velocity

5.1. Distributions of the skewness factor of u′

Figure 15a shows the distributions of the skewness factor $S_{u'} = \overline{u'^3}/\overline{u'^2}^{3/2}$ of the streamwise fluctuating signal in the MBL and SBL versus the vertical distance from the wall, together with results reported by Ueda and Hinze (1975) concerning the SBL. It is seen that the measurements presented here in SBL correspond quite well to Ueda & Hinze (1975) showing the good quality of the measurements near the wall together with some data of in SBL and MBL. The skewness factor $S_{u'}$ presented on Fig. 15a is found to be systematically larger in the MBL than in the standard boundary layer. This observation agrees with Choi's (1988, 1990) and strengthens the idea that the riblets affect especially the higher moments. The increase of $S_{u'}$ in Fig. 15a is about 20% in the MBL, indicating that the degree of asymmetry of the probability density of u' increases by the presence of riblets.

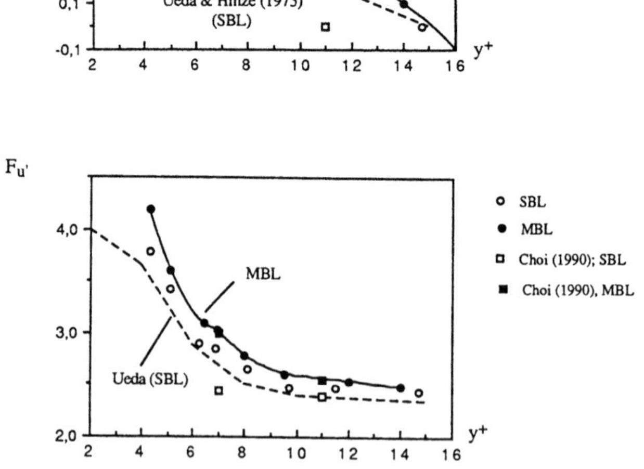

Fig. 15. (a) Distributions of the skewness factor of the streamwise fluctuating signal; (b) Distributions of the flatness factor of u' in MBL and SBL versus y^+.

5.2. Distributions of the flatness factor of the streamwise fluctuating velocity

The effect on the flatness factor $F_{u'} = \overline{u'^4}/\overline{u'^2}^2$ becomes progressively important near the wall (Fig. 15b), where a 12% increase is noted. $F_{u'(\mathrm{MBL})}$ is systematically larger than $F_{u'(\mathrm{SBL})}$ when $y^+ < 15$. Although the flatness factor is not directly related to the intermittency γ, there is a close relationship between $F_{u'}$ and γ and it may be argued that the structure of the internally manipulated layer has a more intermittent character. This again in agreement with Choi (1988, 1990).

6. Statistics of the time derivative of the fluctuating streamwise velocity

6.1. The Taylor and Liepman scales

A 32 point Finite Impulse Response Digital Filter is used to compute the velocity time derivative. It is known that the signal to noise ratio has an important effect on the computation of the velocity derivatives (Kuo and Corrsin, 1971; Ueda and Hinze, 1975). Since the sampling frequency is high compared to the Kolmogoroff frequency here, the cut-off frequency of the digital differentiator was set equal to the Kolmogoroff frequency computed from the data given by Ueda and Hinze (1975), namely to $f_k^+ = U^+/2\pi\eta^+ (\eta^+ \approx 3$ at $y^+ < 10)$. The same configuration of the digital differentiator is used when the computation of the velocity time derivative is computed in SBL and MBL.

The Taylor time scale is defined as:

$$\lambda = \sqrt{\overline{u'^2}/(\overline{\mathrm{d}u'/\mathrm{d}t^2})}$$

and it is related to the viscous dissipation by:

$$\bar{\varepsilon} = 15\nu\left(1 + \frac{\overline{u'^2}}{\bar{u}^2}\right)\Big/\lambda^2$$

a result which is obtained by using Taylor's hypothesis and assuming approximate homogeneity and local isotropy (Ueda and Hinze, 1975). On the other hand, in the case of a gaussian signal, λ is related to the zero-crossing frequency f_0 by:

$$\lambda = \Lambda = (2\pi f_0)^{-1} \quad \text{(Rice, 1945)}$$

where Λ is the Liepman scale. The near wall turbulence has an interesting characteristic because, despite its strongly non-gaussian character, the Taylor and Liepman scales are found almost equal (Sreenivasan et al., 1983).

Figure 16a compares the distribution of the Taylor time scale in manipulated and

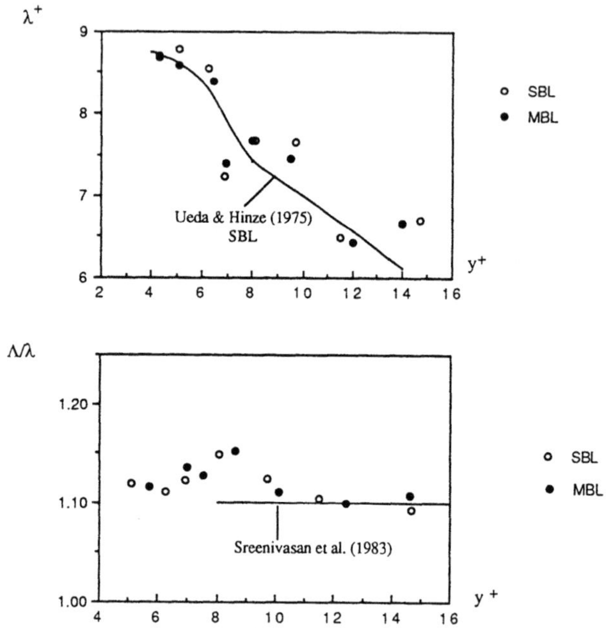

Fig. 16. (a) Distributions of the Taylor time scale in MBL and SBL versus y^+; (b) Ratio of the Liepman time scale to the Taylor time scale in MBL and SBL versus y^+.

standard boundary layer. The agreement with the data of Ueda and Hinze is quite good. There is no effect of the riblets on the Taylor scale nor on the Liepman scale through the whole part of the middle part of the buffer layer. The ratio of the Liepman to the Taylor time scale does not show either any significant modification in the manipulated boundary layer (Fig. 16b). Λ/λ is about 1.1 as is the case in the boundary layer data given by Sreenivasan et al. (1983; Fig. 8a), and this ratio increases slightly near the wall. It may be concluded from the results presented in this paragraph that the Taylor and Liepman time scales are not affected in the MBL, and if the dissipation is altered, it may take place mainly through the differences in the distribution of $\overline{u'^2}/\bar{u}^2$.

6.2. High statistics of the time derivative

The skewness factor of the time derivative $S_{du'/dt}$ is related to the production of vorticity by stretching. Any effect concerning this quantity may imply an alteration of the vorticity production mechanism in the manipulated boundary layer. Fig. 17a compares the $S_{du'/dt}$ distribution in SBL with the $S_{du'/dt}$ distribution in MBL. It is seen that the data presented here agrees well with Ueda and Hinze (1975). $S_{du'/dt}$ is systematically larger in the MBL than in the SBL, at $y^+ < 10$, but this is no longer valid in the beginning part of the buffer layer. Since the variations are small and the trend is not systematic, it is judged that the skewness factor of the velocity derivative is not altered by the presence of the riblets.

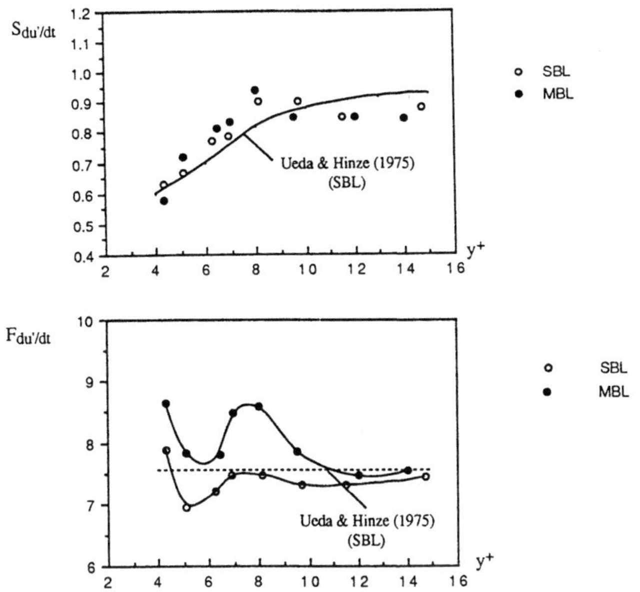

Fig. 17. (a) Distributions of the skewness factor of du'/dt in MBL and SBL versus y^+; (b) Distributions of the flatness factor of du'/dt in MBL and SBL versus y^+.

The flatness factor $F_{du'/dt}$ is slightly Reynolds number dependent, and the data of Ueda and Hinze shown on Fig. 17b was approximately interpolated to $Re_\delta = 27000$ corresponding to the case investigated here. One of the most pronounced effects of the riblets is found in that which concerned with the flatness factor of du'/dt. The increase of $F_{du'/dt}$ is clear and already becomes significant at $y^+ = 12$. One measurement at $y^+ = 14$ and $h^+ = 13.5$ (for which maximum drag reduction is obtained), showed an increase of 15% in the flatness factor of du'/dt, even in the middle part of the buffer layer (Tardu, 1991). These results combined with those concerning $F_{u'}$ show that the riblets have a significant effect on the intermittent structure of the near wall turbulence.

7. Spectra of the streamwise velocity fluctuations

An 8192 point standard FFT procedure was used to compute the spectra of the streamwise velocity fluctuations.

The spectra of the streamwise velocity fluctuations are presented in the same way as Maruyama and Tanaka (1987), in Fig. 18 and the frequency f is non-dimensionalized by the viscous time scale corresponding to the SBL.

The effect on the spectra is confined within $y^+ < 8$. At $y^+ = 4.2$ a decrease in the small frequency range is noted, indicating that the riblets induce a decrease of the large scale structures in connection with the decrease of the ejection and bursting frequency

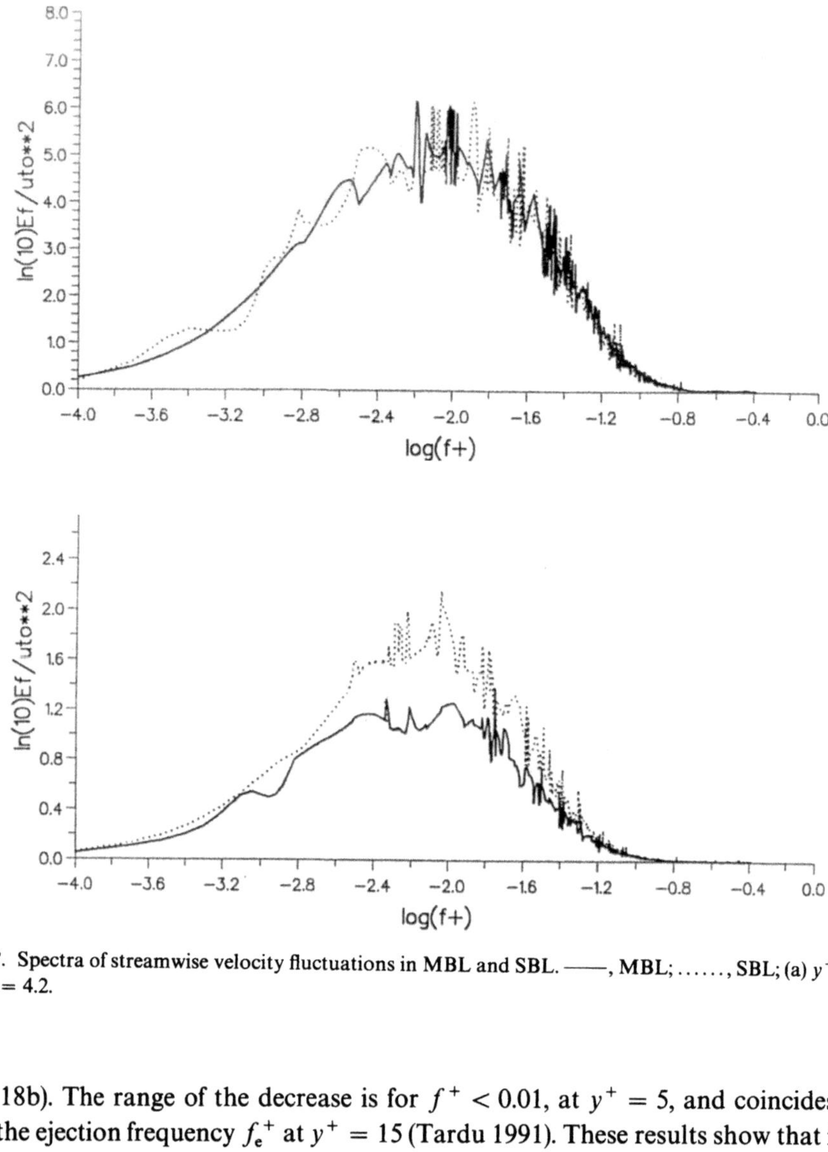

Fig. 18. Spectra of streamwise velocity fluctuations in MBL and SBL. ——, MBL; ......, SBL; (a) $y^+ = 14$; (b) $y^+ = 4.2$.

(Fig. 18b). The range of the decrease is for $f^+ < 0.01$, at $y^+ = 5$, and coincides well with the ejection frequency f_e^+ at $y^+ = 15$ (Tardu 1991). These results show that in the viscous sublayer, a deficit of large scales is induced by the riblets, and this is in agreement with Coustols and Cousteix (1990). In the buffer layer, however, no significant alteration of the small or large scales can be noted from Fig. 18a. Coustols and Cousteix (1990), have reported an excess in the high frequency range, in the region of intense turbulence production, ($y^+ = 10 - 15$), showing that the ribs induce smaller structures there. This point was however not confirmed by the measurements of Choi (1990). Neither do our measurements show an excess in the high frequency range in the part of the buffer layer investigated in this study. The main conclusion of this paragraph is that the effect on the spectra of the fluctuating signal only becomes important in or immediately after the viscous sublayer.

8. Conclusion

A comparative study of the effect of the riblets on the near wall turbulence structure has been carried out. The main conclusions are the following:

(i) The ejection frequency decreases in the middle part of the buffer layer by an amount 3 times larger than the decrease of the inner time scale.
(ii) Both the single and the multiple ejections are affected. The effect on the Multiple Ejection Bursts is more pronounced than on the Single Ejection Bursts. The frequency of the bursts with multiple ejections decreases by 10–20% at the beginning part of the buffer layer. The same behaviour is also observed for the grouped sweeps.
(iii) The skewness factor of the fluctuating streamwise signal increases in the presence of the riblets. The most pronounced effect is found on the flatness factor of both u' and du'/dt indicating an effect of the riblets on the intermittent structure of the near wall turbulence. This is in connection with the decrease of the bursting frequency, which implies a decrease of the intermittency, and consequently an increase of the flatness factor.
(iv) The Taylor and Liepman scales are not affected by the riblets. The spectra of the streamwise fluctuating velocity is modified at the beginning of the viscous sublayer.

It is difficult to explain why the effect on the bursting frequency (by accepting that it is significant) is found only in a limited range of y^+. A possible different redistribution of the $u'v'$ into the bursts may be suspected in the presence of the riblets. More detailed measurements in the buffer layer would therefore be useful. On the other hand, the effect on the spatial extent of the coherent structures has to be investigated. This would clarify how the riblets affect the streak spacing, and how they modify the spatio-temporal distributions of the coherent structures near the wall.

Acknowledgments

This study was carried out while the first author was an invited scientist at the Ecole Polytechnique Fédérale de Lausanne for a period of nine months. He would like to extend his sincere thanks to Prof. I.L. Rhyming for this very kind invitation.

References

Antonia, R.A., Bisset D.K. and Browne L.W.B., Effect of Reynolds number on the topology of organized motion in a Turbulent Boundary Layer. *J. Fluid Mech.* 213 (1990) 267.
Bacher E.V., Smith, C.R., A combined visualization-anemometry study of the turbulent drag reducing mechanism of triangular micro-grove surface modifications. *AIAA* paper AIAA-85-0548 (1985).
Bechert, D.W., Bartenwerfer, M. and Hoppe, G., In: *Proc. 15th ICAS*, London (1986).
Blackwelder, R.F. and Kaplan, R.E., *J. Fluid Mech.* 76 (1976) 89.

Blackwelder, R.F. and Haritonidis, J.H., Scaling of the bursting frequency in turbulent boundary layers. *F. Fluid Mech.* 132 (1983) 87.

Bogard, D.G., Ph.D. Thesis, Purdue University (1982).

Bogard, D.G. and Tiederman, W.G., Burst detection with single point measurements. *J. Fluid Mech.* 162 (1986) 389.

Bogard, D.G. and Tiederman, W.G., Characteristics of ejections in turbulent channel flow. *J. Fluid Mech.* 179 (1987) 179.

Bogard, D.G. and Coughran, M.T., Bursts and ejections in a LEBU-modified boundary layer. In: *Seventh Symposium on Turbulent Shear Flows*, [Toulouse, France, 1987.]

Choi, K.S., A new look at the near-wall structure. In: Matthieu, J. and Compte-Bellot, G. (eds), Berlin: Springer-Verlag (1986).

Choi, K.S., The wall pressure fluctuations of modified turbulent boundary layer with riblets. In: Liepmann H.W. and Narasimha, R. (eds), *Turbulence Management and Relaminarization*, Berlin: Springer-Verlag (1988).

Choi, K.-S., Near-wall structure of a turbulent boundary layer with riblets. *J. Fluid Mech.* 298 (1989) 417.

Choi, K.-S., Effects of longitudinal pressure gradients on turbulent drag reduction with riblets. In: Coustols, E. (ed.), *Turbulence Control by Passive Means*. Dordrecht: Kluwer Academic Publishers (1990).

Coustols, E., Behaviour of internal manipulators: riblet models in subsonic and transonic flows. *AIAA* Paper 89-0963 (1989).

Coustols, E. and Cousteix, J., Experimental investigation of turbulent boundary layers manipulated with internal devices: riblets. In: Gyr, A. (ed.), *Structure of Turbulence and Drag Reduction*. Berlin: Springer-Verlag (1990).

Fulachier, L., Djenidi, L. and Anselmet, F., Couches Limites sur Parois Rainurées Longitudinalement, riblets. 24 ème Colloque d'Aérodynamique Appliquée, Poitiers (1987).

Gallagher, J.A. and Thomas, S.W., Turbulent boundary layer characteristics over streamwise grooves. *AIAA* Paper 84-2185 (1984).

Gaudet, L., Properties of riblets at supersonic speed. *Appl. Sci. Res.* (1989) 245–254.

Kim, H.T., Kline, S.J. and Reynolds, W.C., The production of turbulence near a smooth wall in a turbulent boundary layer. *J. Fluid Mech.* 50 (1971).

Klewicki, J.C. and Falco, R.E., On accurately measuring statistics associated with small-scale structure in turbulent boundary layers using hot-wire probes. *J. Fluid Mech.* 219 (1990) 119.

Kline, S.J. and Robinson, S.K., Turbulent boundary layer structure: progress, status, and challenges. In: Gyr, A. (ed.), *Structure of Turbulence and Drag Reduction*. Berlin: Springer-Verlag (1990).

Kuo, A.Y.S. and Corrsin, S., *J. Fluid Mech.* 50 (1971) 285.

Luchik, T.S. and Tiederman, W.G., Timescale and structure of ejections and bursts in turbulent channel flows. *J. Fluid Mech.* 174 (1987) 529.

Hooshmand, D., Youngs, R. and Wallace, J.M., An experimental study of changes in the structure of a turbulent boundary layer due to surface geometry changes. *AIAA* paper AIAA-83-0230 (1983).

Maruyama, S. and Tanaka, H., The effect of spatial restriction on the inner-layer structure of wall turbulence. *J. Fluid Mech.* 177 (1987) 485.

Offen, G.R. and Kline, S.J., A proposed model of the bursting process in turbulent boundary layers. *J. Fluid Mech.* 62 (1975) 2.

Rohr, J.J., Reidy, L.W. and Anderson, G.W. In: *Drag Reduction 89* Davos: Ellis Harwood Ltd. (1989).

Pulles, C.J.A., Krishna Prasad, K. and Nieuwstadt, F.T.M., Turbulence measurements over longitudinal micro-grooved surfaces. *Appl. Sc. Res.* 46 (1989) 197–208.

Pulles, C.J.A., Krishna Prasad K. and Nieuwstadt, F.T.M., Simultaneous flow visualization and LDA studies over longitudinal micro-grooved surfaces. In: Coustols, E. (ed.), *Turbulence Control by Passive Means*. Dordrecht: Kluwer Academic Publishers (1990).

Pulvin, Ph., Contribution à l'étude des parois rainurées pour les écoulements internes avec gradient de pression positif. Thèse de Doctorat; EPFL, Lausanne, No. 809 (1989).

Rice, S.O., Mathematical analysis of random noise. *Bell Syst. Tech. J.*, 24 (1945) 46.

Robinson, S.K., Kline, S.J. and Spalart, P.R., Quasi-coherent structures in the turbulent boundary layer: Part II. Verification and new information from a numerically simulated flat plate layer. In: *Near Wall Turbulence:* [1988 Zaric Memorial Conference]. Hemisphere (1988).

Robinson, S.K., The kinematics of turbulent boundary layer structure. NASA Technical Memo. 103859 (1991).

Savill, A.M., Effect on turbulent boundary layer structure of longitudinal riblets alone and in combination with outer devices. In: Charnay, L. (ed.), *Flow Visualizations IV.* Hemisphere (1987).
Savill, A.M., Drag reduction by passive devices, a review of some recent developments. In: Gyr, A. (ed.), *Structure of Turbulence and Drag Reduction.* Berlin: Springer-Verlag (1990).
Schwarz-van Manen, A.D., Thijssen, J.H.H., Nieuwvelt, C., Krishna Prasad, K. and Nieuwstad, F.T.M., The bursting process over drag reduction grooved surfaces. In: Gyr, A. (ed.), *Structure of Turbulence and Drag Reduction.* Berlin: Springer-Verlag (1990).
Sreenivasan, K.R., Prabhu, A. and Narasimha, R., Zero-crossings in turbulent signals. *J. Fluid Mech.* 137 (1983) 137.
Tardu, S., Ecoulements Instationnaires en Canal; Réponse des Structures Cohérentes. Ph.D. Thesis; Université J. Fourier-Grenoble-1 (1988).
Tardu, S. and Binder, G., In: *Proceedings of 7th Turbulent Shear Flows.* [1989, Stanford.] (1989).
Tardu, S. and Binder, G., Response of bursting to imposed velocity oscillations. Submitted to *J. Fluid Mech.*
Tardu, S. and Binder, G., Review: effect of the OLDs on near wall coherent structures; discussion and need for future work. In: Choi, K.-S. (ed.), *Recent Developments in Turbulence Management.* Dordrecht: Kluwer Academic Publishers (1991).
Tardu, S., Investigation of the Structure of the Turbulence in an Internal Flow Manipulated by Riblets. Report IMHEF T-91-19; Swiss Federal Institute of Technology, Lausanne; 1015, Lausanne, Switzerland (1991).
Tardu, S., Binder, G., A new method to identify bursting events with single point measurements. In: *Eddy Structure Identification in Free Turbulent Shear Flows* [IUTAM Symposium, 12–14 October 1992, Poitiers, France], (1992).
Tiederman, W.G., Eulerian detection of turbulent bursts. In: *Proceedings of Zaric International Seminar on Near Wall Turbulence* [16–20 May 1988], (1988).
Truong, T.V. and Pulvin, Ph., Influence of wall riblets on diffuser flow. *Appl. Sci. Res.* 46 (1989) 217–227.
Ueda, H. and Hinze, J.O., Fine-structure turbulence in the wall region of a turbulent boundary layer. *J. Fluid Mech.* 67 (1975) 67.
Wallace, J.M. and Balint, J.L., Viscous drag reduction using streamwise aligned riblets: survey and new results. In: Liepmann, H.W. and Narasimha, R. (eds), *Turbulence Management and Relaminarization.* Berlin: Springer-Verlag (1988).
Walsh, M.J., Turbulent boundary layer drag reduction using riblets. *AIAA* Paper AIAA-82-0169 (1982).
Walsh, M.J. and Anders, J.B., Jr., Riblet/LEBU research at NASA Langley. *Appl. Sci. Res.* 46 (1989) 255–262.
Wilmarth, W.W. and Sharma, L.K., Study of turbulent structure with hot wires smaller than viscous length. *J. Fluid Mech.* 142 (1984) 142.

Applied Scientific Research **50:** 215–232, 1993
K. Krishna Prasad (ed.), Further Developments in Turbulence Management.

On near-wall turbulence-generating events in a turbulent boundary layer on a riblet surface

Y.P. TANG[1] & D.G. CLARK*

*Department of Aeronautical Engineering, Queen Mary & Westfield College, (University of London), Mile End Road, London E1 4NS, U.K. (*author for correspondence)*

Abstract. The boundary layer over a drag reducing riblet surface is investigated using hot-wire anemometry and flow visualisation. The concept of a 'riblet sublayer' is introduced, and a definition is proposed in terms of a region of reduced turbulence energy production formed near the wall by the addition of riblets. The hot wire records are examined using a modified form of quadrant analysis, and results obtained over plain and riblet surfaces are compared. Close to the wall, the addition of riblets produces a marked reduction in the occurrence of 'ejection' (2nd quadrant) events. A corresponding increase in the incidence of 'sweep' (4th quadrant) events is accompanied by the development of a strong tendency toward a preferred event duration, and a preferred interval between events. These changes diminish rapidly with distance from the surface, becoming almost undetectable beyond $y^+ = 40$. They are discussed in the light of flow visualisation results, and interpreted in terms of mechanisms associated with the interaction between the riblets and the inner boundary layer flow structures. A conceptual model of the flow mechanisms in the riblet sublayer is proposed.

Key words: Turbulent boundary layer, drag reduction, riblets.

1. Introduction

Since the discovery of 'riblets' as one of the successful passive drag reduction devices investigated at the NASA Langley Research Centre towards the end of the 1970's, extensive investigations have been carried out on both their practical application and the mechanism by which they produce their effect. The latter studies provide information on the manner in which the turbulent boundary layer responds to the disturbance introduced by riblets, and thereby reveal further details about its structure and potential for control. They have therefore been of considerable academic as well as technological interest during the past decade.

Among the investigations which concentrate on the mechanisms of riblet action, that of Gallagher and Thomas (1984) suggested that riblets appear to work by effectively increasing the thickness of the viscous sublayer. Bacher and Smith (1985) considered the interaction of the counter-rotating longitudinal vortices with the small secondary eddies formed at the peaks of the riblets, arguing that the secondary eddies would act to weaken the longitudinal vortices associated with low-speed streaks, as well as to retain low-speed fluid in the grooves. Choi (1989) argued that there is more than one mechanism involved, but that the prime one is the restriction of spanwise movement of the longitudinal vortices, causing a premature and consequently weaker

---

[1]Present address: Dept. of Jet Propulsion, Beijing University of Aeronautics and Astronautics, P.R.C.

near-wall burst. The turbulent wall skin friction is reduced when this happens since, in Choi's view, the near-wall burst is one of the main sources for its production. Walsh (1990) reviewed the various mechanisms which had been proposed to explain drag reduction by riblets. He wrote 'The valleys of narrow v-groove riblets contain low-speed quiescent fluid associated with low skin friction, which lowers the overall drag penalty of the riblets. As the height and spacing of the riblets increases, in terms of wall units, drag reduction is lost. The data indicate that the riblets are still providing lateral resistance to the vortical structures but that more surface area is exposed to high skin friction and the drag penalty has exceeded the benefit.'

A common feature of previous work is that the direct effects of riblets appear to be confined to a viscous flow layer built up in the wall region by the riblet's action, and extending from the surface to $y^+ \approx 18$. It seems reasonable to call this layer the 'riblet viscous sublayer', or just the 'riblet sublayer', although its structure may be very different from that of the sublayer on a plain surface. It is the flow in this riblet sublayer and its interaction with outer flow structures which is likely to be the key to the detailed mechanisms of drag reduction. Unfortunately, for reasons of scale, the flow structure of the riblet sublayer has not been studied in detail.

The present paper is concerned with the statistics of occurrence, and the nature of 'active' flow patterns in the riblet sublayer, and with the interaction between the riblet sublayer and outer flow structure. In this context, an 'active' flow pattern is one associated with a marked excursion of the velocity perturbation into the second or fourth quadrant of the $u' - v'$ plane. Such patterns account for the majority of turbulence energy generation in the layer, and their passage is detected by the perturbation 'event' they produce in the output from a stationary hot wire probe. Such events are referred to below as 'active events', but it must be emphasised that the occurrence of the 'event' does not necessarily imply any particularly rapid evolution of the associated flow pattern, but simply its passage downstream past the probe at the local convection velocity.

In the present investigation, a modified quadrant analysis (MQA) scheme was used for detection and data analysis. With this technique, some distinctive characteristics of active events have been found in the riblet sublayer, and some aspects of the associated flow structure have been examined with the help of the flow visualisation. A conceptual model of the riblet viscous sublayer has been proposed from the experimental findings.

2. Experimental arrangement and procedure

The experiments were conducted in the fully developed turbulent boundary layer on the floor of a low-speed wind tunnel, with the adjustable working section roof set to give zero longitudinal pressure gradient. The tunnel free stream turbulence level was approximately 0.5% under these conditions. Further details of the facility and methods used, have been given by Clark (1990), and Tang and Clark (1991).

Two L-shape riblet surfaces were used. The first has dimensions $h = 3$ mm, $s = 4$ mm and $l = 1$ m, corresponding to $h^+ = 15$, $s^+ = 20$ and $l^+ = 5000$ at $R_\theta = 1400$. The second has the dimensions $h = 5$ mm, $s = 10$ mm and $l = 1$ m, corresponding to $h^+ = 12.5$, $s^+ = 25$ and $l^+ = 2500$ at $R_\theta = 650$. The two cross-sections are shown in Fig. 1. The riblets consist of strips of black polystyrene, glued to the surface with polystyrene adhesive, and subsequently shaped by scraping with a form cutter. In both cases, the underlying surface consists of a sheet of 6 mm plate glass, with a layer of matte-black paint on its upper face. The light planes for flow visualisation are formed by placing a line of 12 volt, cine-projector lamps beneath 1.5 mm slits cut in the paint layer.

Velocity measurement in the wind tunnel was made using a DISA 55P61 x-array hot wire probe. This probe has two wires 5 μm in diameter and 1.25 mm in length, with a spacing of 0.5 mm between the wires. When normalised with inner variables at the highest Reynolds number of the present experiments, this gives a length of 6 wall units and a spacing of 2.5 wall units. Two DISA 55M10 units were used to operate the probes. A square-wave test gave a frequency response of 30 kHz for the system with an over-heat ratio of 1.35.

The probe was directly calibrated at the centreline of the tunnel, over the range $-18°$ to $+18°$, at five different velocities. A DANTEC Precision Anemometer 54N60 was used to establish the tunnel centreline velocity. Effects due to spanwise velocity perturbations were neglected, the estimated maximum error due to this cause being less than 3.5% for streamwise velocity and negligible for normal-velocity measurements.

A High Performance Data Acquisition Card was installed in a PC-AT-286 personal computer to digitise the two channels of hot wire signals with 12 bit accuracy.

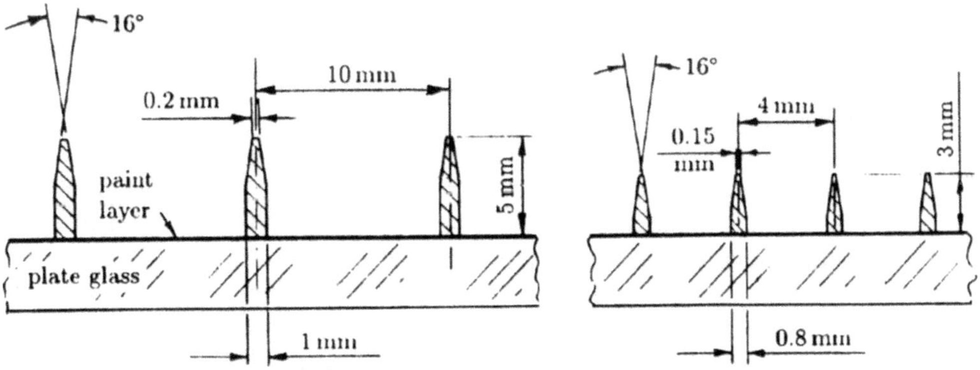

Fig. 1. Riblet dimensions.

For flow visualisation, smoke was introduced using a smoke wire 12 μm in diameter. The estimated maximum Reynolds number of the wire was 1, consequently, no significant disturbance was anticipated from the wire itself. Single or multiple transverse light planes, sloping upstream from the surface, were used for illumination, the resulting sections being viewed from downstream in a direction approximately normal to the light planes (Clark 1990). Using multiple light planes, this technique provides a useful level of three dimensional information.

The visualisation studies were viewed, recorded and evaluated using a standard video system comprising a video camera (Panasonic WV-F15), a 14 in. colour monitor, and a high quality video recorder (Sony VO-9600P). A video sync separator chip, supplied with the video camera signal, was used to output synchronisation pulses to the computer data acquisition system in order to establish the precise relationship between the hot wire signal and the flow visualisation images. With the PAL interleaved line video system, the standard television framing rate of 25 full picture frames per second gives 50 field scans per second at half the full number of 625 lines per frame. The field scans can be viewed separately by the slow moving and freeze frame facility on a video recorder giving an effective framing rate of 50 fps. In the present work, this gave four or five frames during the average duration of the most active events at $R_\theta = 750$.

The basic scheme of the present experiment follows the observation of Bogard and Tiederman (1986), that hot wire measurement is effective in detecting 'events', whereas flow visualisation is effective in viewing the associated flow structure. The experiments were conducted in two phases. In the first phase, hot wire records were obtained on the riblet surfaces at $R_\theta = 1400$ without flow visualisation. The aims of this phase were as follows; (i) to compare the mean flow characteristics (mean velocity, turbulence intensity, turbulence shear stress and spectrum) between the plain and riblet surfaces, (ii) to compare the statistics of the active events on the plain and riblet surfaces detected by the MQA technique.

In the second phase, hot wire records were obtained at $R_\theta = 650$ with simultaneous flow visualisation, in order to identify flow structures associated with important active events in the hot wire record. Some calibration runs were also carried out without the flow visualisation, to check for changes in the perturbation statistics which could be attributed to effects of heat from the smoke wire or the lighting on the hot wire. No significant changes were found.

A basic data group was collected from phase one consisting of the statistics of occurrence of active events at four distances from the surface; $y^+ = 18$, 30, 40 and 60, over both plain and riblet surfaces. A recording time of 240 s was used at each position, with a sampling rate of 500/s per channel. This corresponds to an interval between data points of $t^+ = 0.74$ and gives at least five data points in the duration of an interesting event. Several runs were made at each probe position, and no significant difference in the statistics was found between different runs. The statistics quoted below are all obtained from these data groups.

3. Data reduction

In the modified quadrant analysis (MQA) technique used in the data reduction, an active event is defined as one in which the perturbation product $u'v'$ is negative and has an absolute value above a certain threshold level. The individual events are then placed in categories depending on the maximum value of $|u'v'|$ reached, and whether the excursion is into the second or fourth quadrant of the $u' - v'$ plane. Further categories, depending on the nature of the trajectory in the $u' - v'$ plane, have also been examined (Tang and Clark (1991)), but are not used here.

Formally then:

(i) An active event is one associated with a velocity perturbation which makes a positive contribution to turbulence production, that is, the perturbation lies in the second or fourth quadrant of the $u' - v'$ plane. In addition, the absolute value of the perturbation product $u'v'$ must rise above a certain threshold value.

(ii) Threshold levels Th_n, are defined as follows;

$$\text{putting} \quad Th = u'_{\text{rms}} v'_{\text{rms}},$$

$$Th_n = n \times Th \quad \text{for} \quad n = 1, 2, 3, \text{ and } 4.$$

(iii) An event starts when $|u'v'|$ rises above the threshold Th_1 and ends when $|u'v'|$ falls below it.

(iv) Events are classified according to the highest threshold crossed by the value of $|u'v'|$. A Th1 event is one in which the maximum value lies between Th_1 and Th_2; a Th2 event, one in which the maximum reached lies between Th_2 and Th_3, and so on, to Th4. Events which would fall into categories beyond Th4 are not sufficiently numerous for their further classification to be statistically significant, hence effectively, $Th_5 = \infty$.

(v) Events for which v' is negative are characterised as 'sweeps', and events for which v' is positive are characterised as 'ejections'.

The time interval t_i between the peak values of two consecutive events and the duration t_d of individual active events are defined as shown in Fig. 2. Representative ejection and sweep events are shown in Fig. 3 as trajectories in the $u' - v'$ plane.

Results

The mean-velocity profiles measured at $R_\theta = 1400$ on the plain and riblet surfaces are presented in Fig. 4. A single hot-wire probe was used, and a correction for wall proximity was applied to those points obtained at less than 500 wire diameters from the wall.

For the plain surface, the logarithmic region is well described using a von Karman

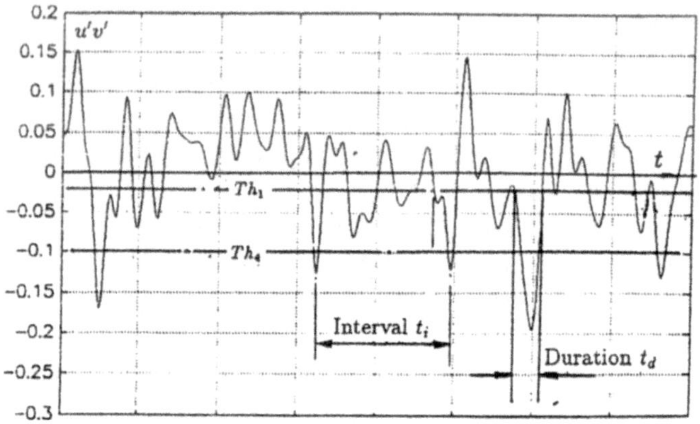

Fig. 2. Definition of interval and duration.

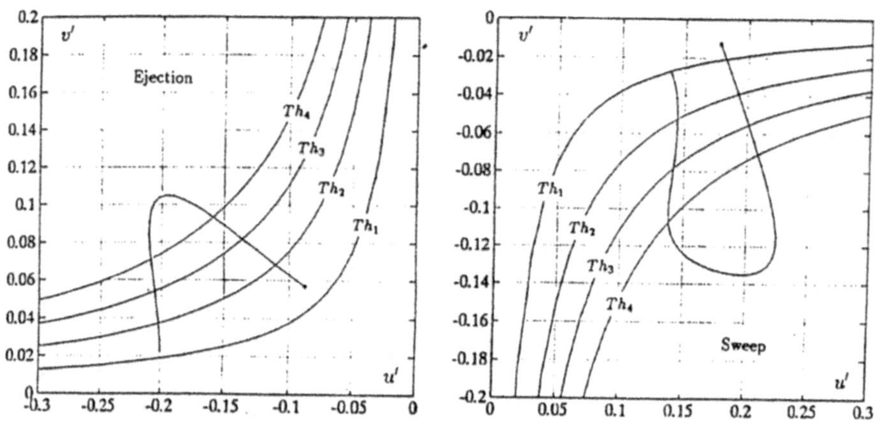

Fig. 3. Representative examples of event trajectories in the $u' - v'$ plane.

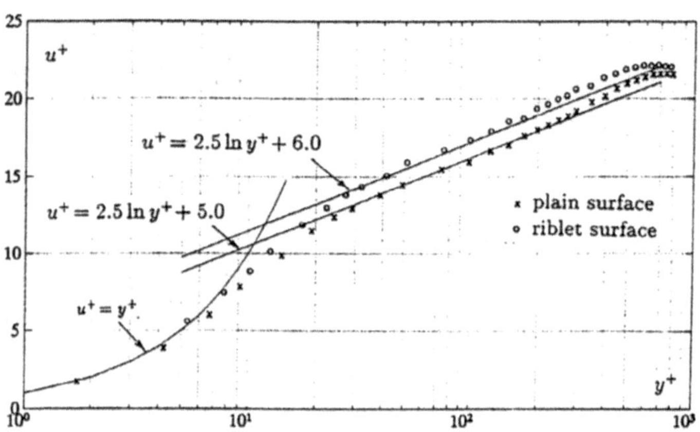

Fig. 4. Mean velocity over plain and riblet surfaces.

constant of 0.4 and a logarithmic intercept of 5.0. Estimates of the friction velocity u_τ, were made using both a Clauser plot and an estimate of the velocity gradient at the wall. The estimates agreed within 2%, and the mean of the two values was used in subsequent data reduction. The value of u_τ, the wall shear stress coefficient c_f, and shape factor H are shown in Table 1. The size of the corresponding wall units for length and time, y_τ and t_τ, are also given. For the riblet surface, the friction velocity u_τ and the virtual origin y_v were determined by Choi's method (Choi 1989), and the resulting values were used to obtain the points plotted in Fig. 4. These points indicate a logarithmic region extending from $y^+ = 35$ to about $y^+ = 200$, with a slope, $1/k$, unchanged from that on the plain surface, but with an intercept at $y = 0$ increased from 5.0 to 6.0; a shift away from the surface which is consistent with the results of other studies (e.g. Choi (1989)). The friction velocity u_τ is 2.5% lower than that on the plain surface, corresponding to a 5% reduction of shear stress at the wall, with $y_v = 2.4\,\text{mm}$, that is $y_v^+ = 12$. The turbulence intensity of the u velocity component and the distribution of turbulence shear stress $\overline{u'v'}$ are shown in Fig. 5, (a) and (b). Both figures cover the range $y^+ \leqslant 200$, to include the full logarithmic region. The origin in both these plots is the surface of the basic plate, and the variables are all reduced to non-dimensional form using the friction velocity u_τ on the plain surface. In Fig. 5(a) the peak in the u' profile shifts from $y^+ = 15$ on the plain surface to $y^+ = 25$ over the riblet surface, and the maximum value is reduced by 7%. However, beyond

Table 1. Boundary layer parameters for the plain surface.

R_θ	u_τ m/s²	c_f	H	y_τ mm	t_τ s
1400	0.0742	0.00417	1.41	0.196	0.0028

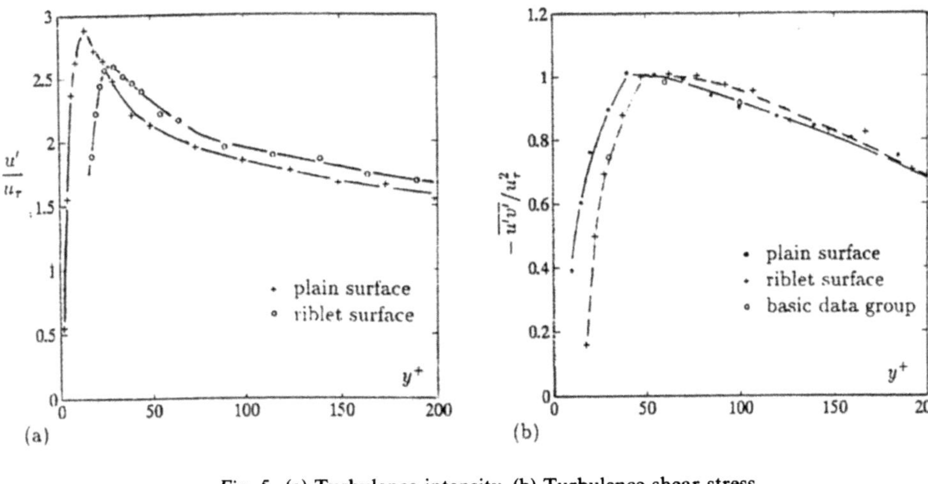

Fig. 5. (a) Turbulence intensity. (b) Turbulence shear stress.

about $y^+ = 30$ the value of u' on the riblet surface is the higher of the two. Introducing a virtual origin for the riblet surface results would bring the two curves into close agreement except in the region of the peak between $y^+ \approx 10$ and $y^+ \approx 35$. In Fig. 5(b) the $\overline{u'v'}$ profile on the riblet surface shows features similar to those of the u' profile. No obvious decrease of shear stress is apparent.

The turbulence spectra for the riblet surface at $y^+ = 18$ and the plain surface at $y^+ = 6$ are shown in Fig. 6(a). Two lines with slopes of -1 and $-5/3$ are also shown for comparison. The spectrum over the plain surface does have identifiable linear regions parallel to these lines, but in the case of the riblet surface, the region with a slope of -1 is absent. The results suggest that the addition of riblets produces a significant decrease of energy-containing turbulence for wave numbers less than 0.02, and some increase for wave numbers between 0.02 and 0.2. The absence of the region with a slope of -1 is associated with this change.

Perry et al. (1986) pointed out the relationship between the turbulence spectrum and the presence of coherent flow structures, and reconstructed a spectrum with a slope of -1 by generating artificially the type of signal to be expected from the passage of a probe through vortex-like structures. This suggests that the change in the turbulence spectrum may result from a significant change in the flow structure when riblets are added to the surface.

The distribution of turbulence production from Figs 4 and 5(b) is shown in Fig. 6(b). For the purposes of discussion below, the boundary layer is divided into three regions based on these results. In region I ($y^+ <\approx 19$) the turbulence production over the riblet surface decreases rapidly and is apparently lower than that over the plain surface. In region II ($19 < y^+ < 60$), the turbulence production over the riblet surface is higher than that over the plain surface, mainly due to the fact that the local mean flow shear rate is higher. In region III ($y^+ > 60$), there are no significant differences between the plain and riblet surfaces.

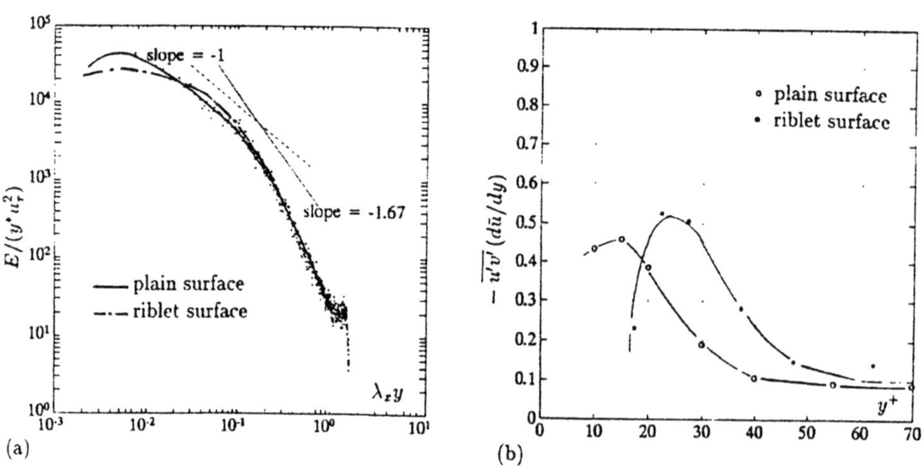

Fig. 6. (a) Turbulence spectra. (b) Turbulence production.

4.1. Examination of active events in region I

Histograms of the occurrence of active events over the plain and riblet surfaces at $y^+ = 18$ are shown in Fig. 7.

The probability is calculated by:

$$P_{n_j} = N_{n_j}/N, \quad \text{where } n = 1, 2, 3, 4 \quad \text{and } j = 1, 2.$$

Here N is the total number of active events. N_n is the number of events of rank n, and j denotes the type of event; $j = 1$ indicates an ejection ($v' > 0$), and $j = 2$ indicates a sweep ($v' < 0$). The most noticeable change associated with the addition of riblets is a dramatic increase in the number of sweep events. As the rank of the events increases, sweeps become the dominant event and ejection events become very rare.

A study which only included ejections in an assessment a 'bursting' rate, would conclude that riblets cause a marked reduction in bursting rate at this position. If both sweeps and ejections were included, the bursting rate over the riblet surface, (using Th_2 as a threshold) would be about 10% higher than that over the plain surface, but the burst duration would be about 10% lower. In other words, riblets appear to cause a shift toward more sweeps of short duration near the wall.

More details of this distinctive feature of the riblet surface data can be seen in Fig. 8. The threshold (or hole size) used here is Th_2. Figure 8a shows the probability distribution versus time interval for individual sweep events and Fig. 8b shows the probability versus duration. From Fig. 8 the main difference between riblet and plain surfaces occurs when the interval, t_i^+, is less than 200. In this low interval range the riblets create a peak of sweep probability at $t_i^+ = 18$, though the mean interval (averaged between $t_i^+ = 15$ and $t_i^+ = 200$) is about 36.

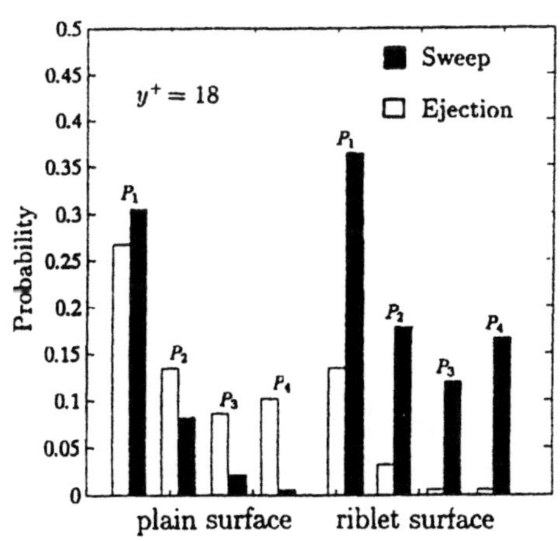

Fig. 7. Event histograms: region I.

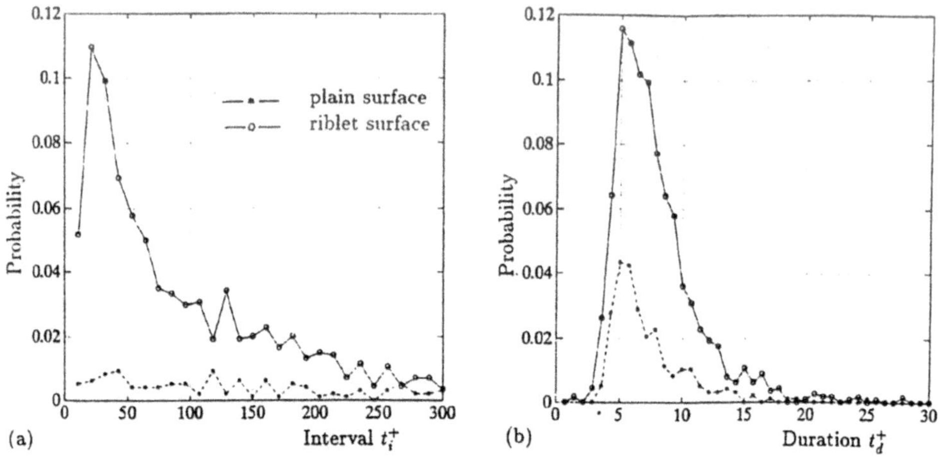

Fig. 8. Probability of a sweep event; region I.

Figure 8b shows that the increase of the probability of sweep events occurs for event durations, t_i^+, of between 3.6 and 14. This is the range of duration for which the probability of a sweep event over the plain surface is also high. The position of the peaks is almost same over the two types of surface and the form of the curves is similar.

Figure 9 shows the probability of ejection events versus interval and duration. Unlike the probability of sweep events, the probability of ejection events in this near-wall region is reduced by the addition of riblets; a point which is also revealed by the histogram of Fig. 7.

The main changes shown in Fig. 7 and Fig. 12, were found to be independent of both the spanwise position of the probe and the threshold (or hole size) used.

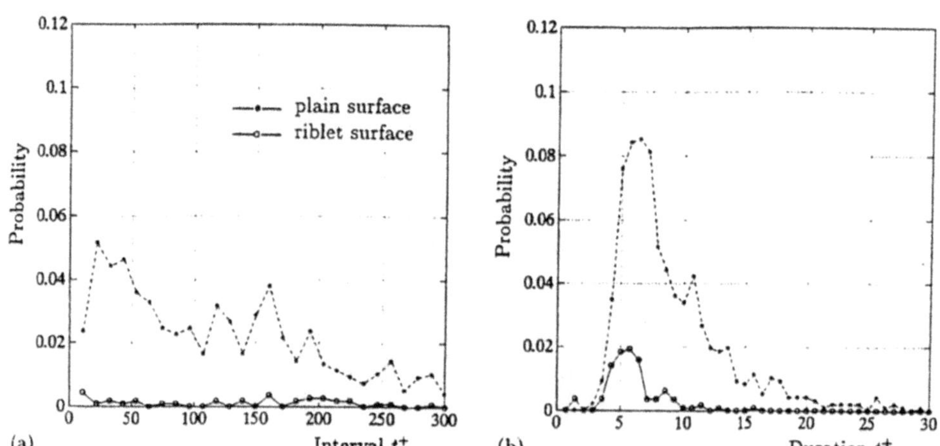

Fig. 9. Probability of an ejection event; region I.

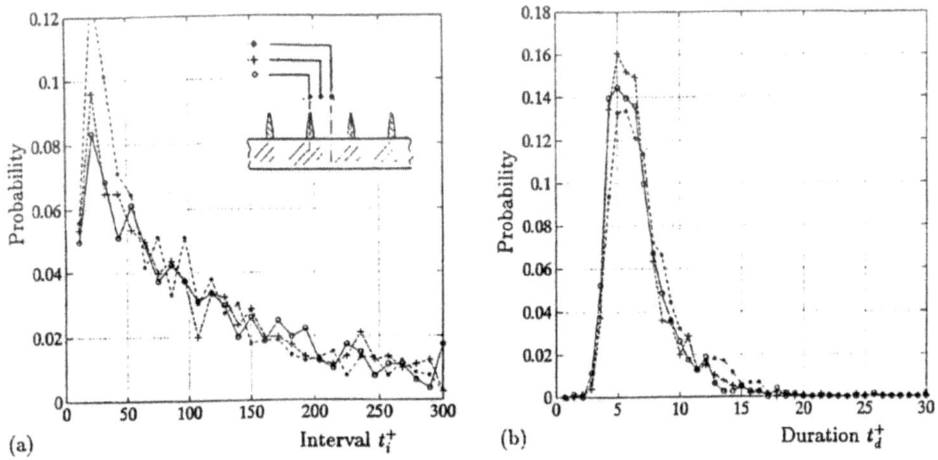

Fig. 10. Probability of a sweep event at three spanwise positions ($y^+ = 18$).

Figure 10 shows the probability of the sweep events versus (a) the interval, and (b) the duration, at $y^+ = 18$, measured at three different spanwise positions above one riblet groove. The basic characteristics, such as the interval corresponding to the peak of probability, are similar at all spanwise positions.

Differences in the amplitude of the probability are within the range of scatter of the results, except for considerable differences in the regions of peak probability.

Figure 11 shows the probability of sweep events versus the interval of individual sweep events at a threshold Th_4. Compared with Fig. 8a the distribution is similar, though the preferred interval has increased to about $t_i^+ = 100$, and naturally, the peak probability is lower.

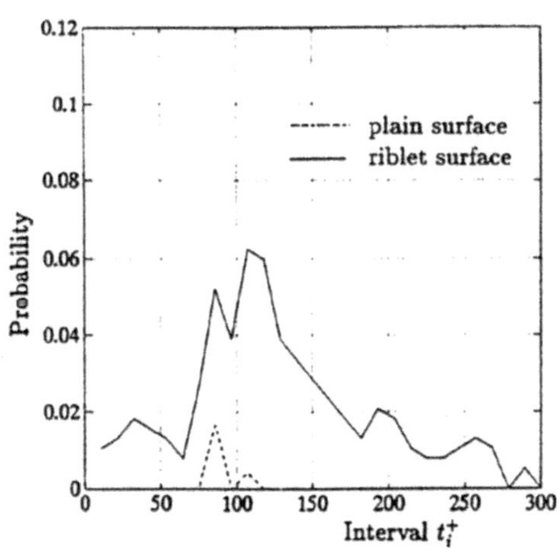

Fig. 11. Probability of Th_4 sweep events; region I.

4.2. Examination of active events in region II

A histogram of active events in region II is shown in Fig. 12. The probability of ejection events recovers rapidly with increasing distance from the surface, though at $y^+ = 30$ it is still lower than that over the plain surface, whilst for higher rank events, the probability of a sweep is still higher. The differences are further reduced at $y^+ = 40$.

Figure 13 shows the probability of sweep or ejection versus interval and duration of events using a Th_2 threshold. The scale of Fig. 13 is the same as that of Fig. 8 for comparison. The distinctive feature in Fig. 13a is that the occurrence of a sweep loses its dominant central inverval and spreads over a wide range of intervals. The further from the wall, the flatter the distribution, until at $y^+ = 40$ it becomes similar to that over the plain surface. That the probability of occurrence of an ejection event increases, as the probe is moved away from the wall, can be seen in Fig. 13b, but the distribution is not substantially different from that over the plain surface.

Comparing the statistics of active events over the plain and riblet surfaces in Fig. 13 and Fig. 8; the probability versus the interval of sweep events over the plain surface retains similar characteristics at all probe positions; the amplitude of the probability is similar and sweeps occur at a wide range of intervals. The probability over the riblet surface has very different characteristics. In region I, there is a sudden increase of the probability of a sweep event occurring with a marked preferred interval. The characteristics return to those of the plain surface as the probe is moved out through region II.

Analysis of the statistical characteristics of ejection events leads to a similar picture. It is reasonable to ascribe these characteristics to the presence of local flow structures with a scale of region I.

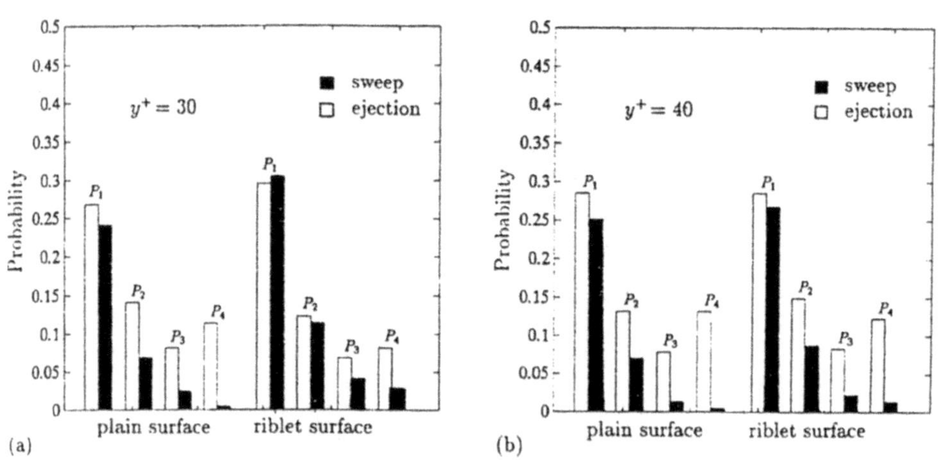

Fig. 12. Event histograms; region II.

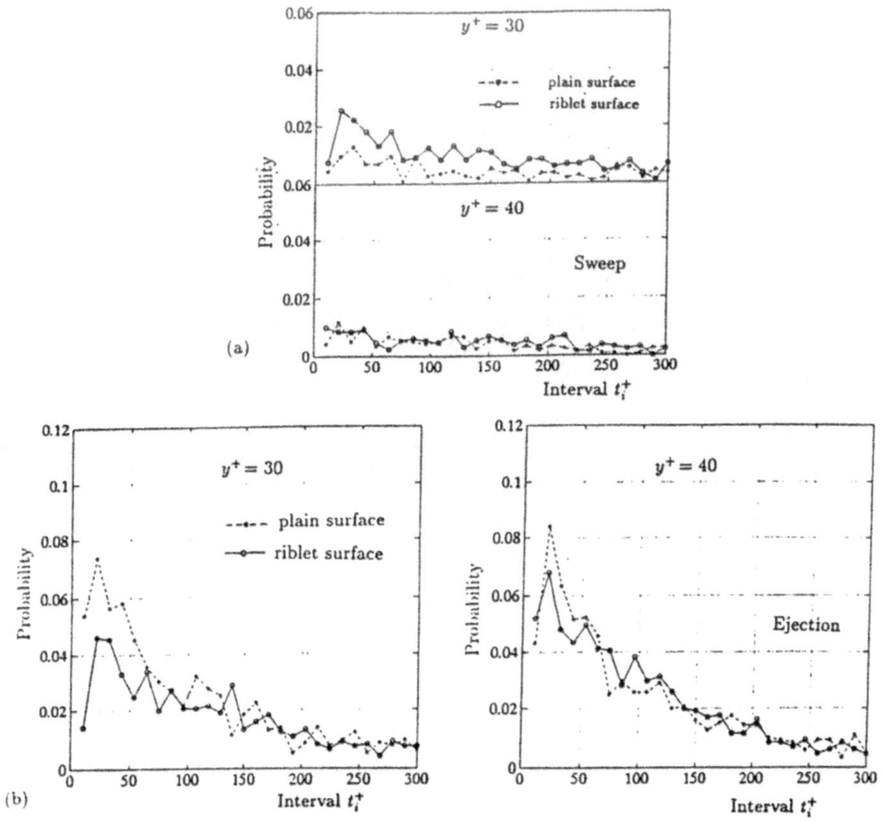

Fig. 13. Probability of sweep and ejection events; region II.

4.3. Flow patterns associated with active events

Flow visualisation on the riblet surface, using both single and multiple light planes reveals very complex flow patterns with different time and space scales. Those associated with MQA events, can be identified with the aid of a simultaneous hot wire record from a probe situated in the field of view, and in spite of the apparent complexity, some characteristic features are apparent.

Two patterns originating in motions outside the riblet sublayer were identified as being associated with strong (Th4) sweep events at $y^+ = 18$. A representative example of the first type is shown in Fig. 14. Two frames of a video sequence are shown. These were obtained using single light plane, and with smoke originating from a wire at the level of the riblet tops. The time interval between the frames is 0.1 s ($t^+ \approx 36$). The second frame shows a local clearance of smoke-contaminated flow from the riblet grooves with no obvious flow structure visible within the cleared region. The clearance of smoke can only be due to a sweep originating from large scale motion in the outer layers, and the patterns are consistent with the occurrence of motion of the

Fig. 14. Flow visualisation during a strong sweep event in region I.

type which is responsible for the formation of a 'Falco pocket' in a near-wall smoke layer on a plain surface. Only by its absence can smoke originating from the inner layers reveal a sweep of this type.

The second example of a flow pattern associated with a strong (Th4) event is that of the interaction of a pair of longitudinal vortices causing a sweep at the position of the probe. In Fig. 15(a) an example of this feature is shown using a single light plane, and in Fig. 15(b) using multiple light planes with simultaneous hot wire measurement. The longitudinal vortices have a scale of $l^+ \approx 30$. As with the previous example, this feature is also found over a plain surface.

Figure 16(a) shows an example of the secondary eddy at the top of a riblet, mentioned by Bacher and Smith (1985). In the case shown, smoke was generated from a streamwise wire positioned along the top of the riblet. On the video record it is clear that the eddy results from crossflow at the riblet top, due to spanwise meandering of

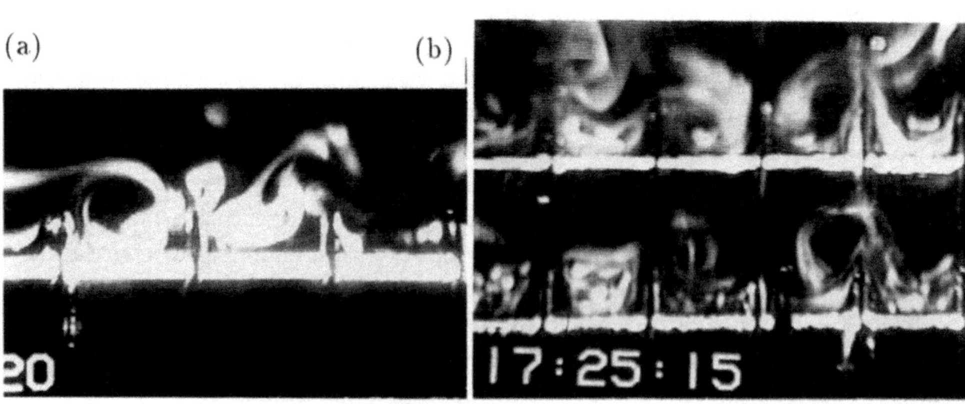

Fig. 15. Strong sweep events in region I; vortex pairs.

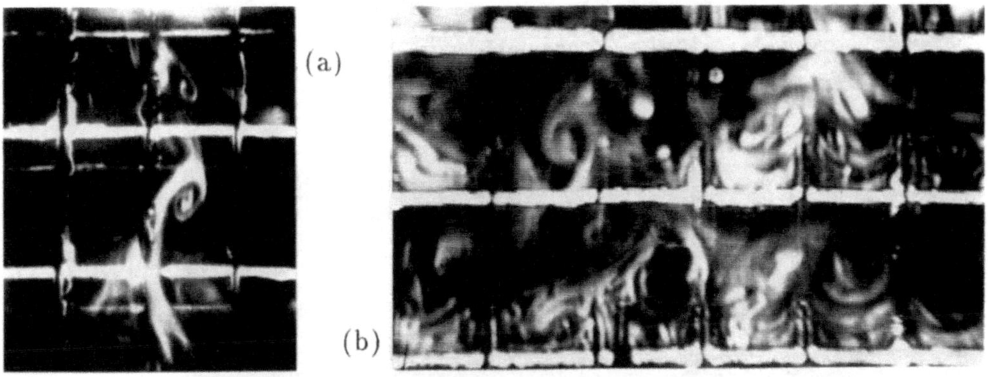

Fig. 16. (a) Secondary eddy. (b) Complex interaction.

the overlying flow. The video records also reveal example of apparent interaction between these secondary eddies and the longitudinal vortices in the outer flow, interactions which are also accompanied by Th4 sweep events at $y^+ = 18$. An example of the complex patterns which occur such cases is shown in Fig. 16(b). In some such sequences, the outer longitudinal vortex appears to be drawn down into the riblet groove. These observations lend support to the suggestion of Bacher and Smith that interactions of this type represent a significant mechanism for weakening the mixing process.

Figure 17 shows a series of smoke patterns illuminated by a single light plane. The smoke wire was positioned at the top of riblets and at a distance $l^+ = 150$ upstream of the light plane. The series covers 0.3 s ($t^+ \approx 108$) and the interval between any two frames is 0.1 s. At time t_0 the smoke initially moves up away from the groove, and then gradually moves right down into the groove, until at t_3 it reaches the bottom. There are no violent local flow patterns associated with this process. The video sequence suggests that the flow is drawn down in this way by the influence of an energetic sweep downstream. This accelerates the flow along the groove locally, causing flow to be drawn into the groove for some distance upstream. If this is the case, a corresponding flow up out of the groove is likely to be present downstream, and behaviour which could be interpreted in this way was also observed, elsewhere in the video record. These processes could be termed respectively, 'riblet induction' or 'riblet suction', and 'riblet overflow'.

The motions noted above are all on the scale of region I and are therefore likely to be the motions responsible for the altered near-wall turbulence statistics on the riblet surface but they all result from processes originating above region I and become part of them. In the log-law region, therefore, their effect is to modify the flow structures without creating fundamentally new ones.

Strong ejections are rare in the near wall region, and it seems that the processes which are accompanied by flow out of the grooves, tend to be gentle and extended, rather than abrupt, localised and violent.

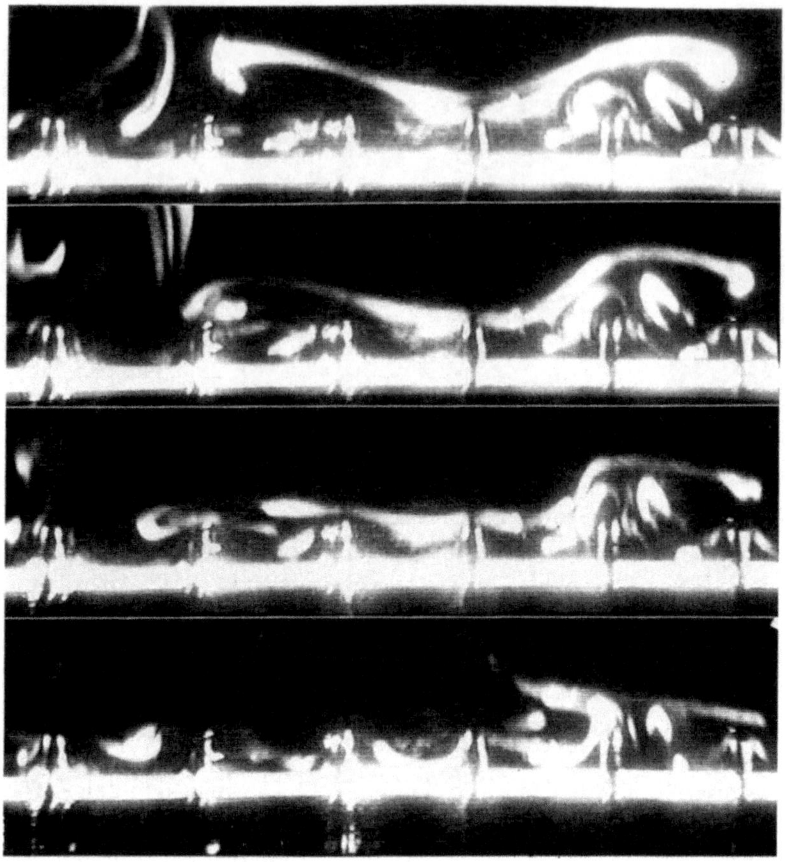

Fig. 17. **'Riblet induction'.**

5. A conceptual model of the riblet sublayer

The results discussed above provide further evidence that on a drag-reducing riblet surface, a region exists which may conveniently be referred to as the 'riblet sublayer'. Within this layer, the characteristics of the flow are significantly altered by the presence of the riblets and some flow features exist which are unique to the layer. These features are small in scale, and consist simply of interactions between the riblets and the flow configurations normally found in the near-wall layers of a turbulent boundary layer. They do not constitute a change in the basic flow mechanisms normally present in the inner log-law region.

In terms of the methods adopted in the present work, the riblet sublayer may be identified with a region in which the presence of the riblets produces a substantial alteration in the distribution of 'active events' in the output of an x-wire anemometer probe aligned to reveal $u' - v'$ perturbations. The layer is characterised by an increase in 'sweep' (4th quadrant) events of short duration, and a marked reduction in 'ejection'

(2nd quadrant) events. The sweep events occur over a wide range of intervals, but have a preferred interval of $t^+ \approx 20$. Strong ejection events become very rare, and no preferred interval between occurrences is discernible. The layer extends from the surface to $y^+ \approx 20$.

Beyond $y^+ \approx 20$, the predominance of 'sweep' events persists to $y^+ \approx 40$, but the preferred interval is no longer present, and the event statistics rapidly return to those found over a plain surface. In those terms, at $y^+ = 60$ the two surfaces are indistinguishable.

Viewing the flow visualisation results over the riblet surface in the light of the event statistics, the flow patterns which are similar to those over a plain surface are the large scale outer-sweep, penetrating to region I, and the medium scale vortex-induced sweep in region II. These two features are also found to occur with a wide range of intervals on a plain surface, and their scale suggests that they can only be involved in the reduction in the number of Th4 events, not the increased number of events of lower rank.

Three types of pattern which could account for the altered event statistics in region I are identified as the small scale secondary eddies forming in sporadic cross-flow across the riblet peaks, the parts of longitudinal near-wall vortices which enter the riblet grooves, and the tendency for sweep flow from the log-layer to entrain further flow into the riblet grooves by mass conservation. The scale of these patterns is consistent with the relatively short interval and duration of the extra sweep events detected.

The foregoing suggests a conceptual model of the riblet viscous sublayer as illustrated in Fig. 18. The flow in the outer layers is taken to be dominated by vortical elements, originating from the evolution of hairpin eddies similar to those proposed by Wallace (1990). Riblets bring about no fundamental change in the large and middle-scale vortical elements but they do change the environment in which such features are initiated. The sublayer which they create is highly directional, and above it where the flow has sufficient freedom for turbulence to be produced, rates of shear

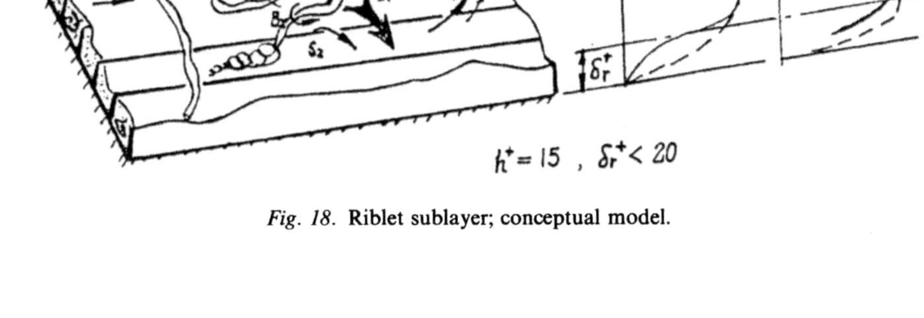

Fig. 18. Riblet sublayer; conceptual model.

are lower than those in the equivalent layer over a plain surface. High speed, highly perturbed fluid swept into the riblet grooves (S1) encounters strong viscous dissipation, and emerges with its perturbation level greatly reduced and any lateral component of velocity largely removed. The increased spacing of the low speed streaks above the riblet surface may be explained by the reduced mean flow shear rate. Interaction between the streamwise vortical structures and the riblets, with some parts of the flow structures being drawn down into the riblet grooves may also contribute. The smaller scale processes of the secondary-eddy induced sweep (S3) and 'riblet induction' (S2) are responsible for less energetic movements of fluid into the grooves and are the features most directly responsible for the altered event statistics characterising region I. The slow moving, quiescent fluid from the grooves emerges, either by the action of an outer vortical structure producing a strong ejection, (E1) or simply by being gradually displaced (E2) as other more energetic fluid enters elsewhere, the latter being the dominant outward flow feature in the sublayer. The flow rate in and out of the riblet grooves should be directly related to the balance of mean momentum and turbulence within them, and will therefore be related to the skin friction.

Acknowledgments

The work was supported by a grant from the UK Department of Trade and Industry.

References

Bacher, E.V. and Smith, C.R., A combined visualisation-anemometer study of the turbulent drag reducing mechanisms of triangular micro-groove surface modification. *AIAA* Paper 85-0548 (1985).

Bogard, D.G. and Tiederman, W.G., Burst detection with single-point velocity measurements. *J. Fluid Mech.* 162 (1986) 389.

Choi, K.-S., Near-wall structure of a turbulent boundary layer with riblets. *J. Fluid Mech.* 208 (1989) 417.

Clark, D.G., Boundary layer flow visualisation patterns on a riblet surface. In: Coustols, E. (ed.), *Turbulence Control by Passive Means.* Dordrecht: Kluwer Academic Publishers (1990).

Gallagher, J.A. and Thomas, A.S.W., Turbulent boundary layer characteristics over streamwise grooves. *AIAA* Paper 84-2185 (1984).

Perry, A.E., Henbest, S. and Chong, M.S., A theoretical and experimental study of wall turbulence. *J. Fluid Mech.* 165 (1986) 163.

Tang, Y.P. and Clark, D.G., On turbulence generating events in turbulent boundary layer. *QMW-EP*-1086 (1991).

Wallace, J.M., On the structure of bounded turbulent shear flows. In: *Proc. NASA Boundary Layer Structure Workshop.* [Langley Research Center, Hampton, Virginia, August 28–30, 1990.]

Walsh, M.J., Riblets. In: Bushnell, D.M. and Hefner, J.N. (eds), *Viscous Drag Reduction in Boundary Layers.* Progress in Astronautics and Aeronautics 123. (1990) p. 203.

Applied Scientific Research **50**: 233–254, 1993.
K. Krishna Prasad (ed.), Further Developments in Turbulence Management.

Friction velocity and virtual origin estimates for mean velocity profiles above smooth and triangular riblet surfaces

A.D. SCHWARZ-VAN MANEN, A.F.M. VAN GELOVEN,
J. NIEUWENHUIZEN, J.C. STOUTHART, K. KRISHNA PRASAD
& F.T.M. NIEUWSTADT*
*Eindhoven University of Technology, The Netherlands; * Delft University of Technology, The Netherlands*

1. Introduction

It is well known that riblet surfaces influence only the region close to the wall (Hooshmand et al. 1983). The appropriate velocity scaling for this region is the friction velocity (u_*). Thus it is important to determine u_* very accurately. For situations where no drag balance is available various methods have been developed to estimate u_* from the mean streamwise velocity profile. Generally, the logarithmic friction law (Tennekes and Lumley, 1972) is used, better known as the log-law or the law of the wall method. However, most of the methods can only be used for surfaces where the $y=0$ position is exactly known, i.e. for smooth surfaces. Choi (1989) developed a method for riblet surfaces to determine from the mean streamwise velocity profile both the friction velocity and the virtual origin of the riblets into the turbulent boundary layer. In that same year, Bechert and Bartenwerfer (1989) published a theoretical model, developed for riblet surfaces to calculate the protrusion height.

During the investigation presented here, it was discovered that the log-law method could not be used; the error in the friction velocity was of the same order as the amount of drag reduction of the investigated riblet surfaces. The two main sources for this error are: (i) the estimation of the logarithmic region in the vertical velocity profile of the mean stream; and (ii) the Von Kàrmàn constant (κ). In the literature values for κ are given between 0.38 and 0.42. For the same reason this method was criticized by Bradshaw et al. (1991). Therefore, several methods were used to estimate the friction velocity for the smooth surface from the mean streamwise velocity profile and the respective results were compared with one another. One of the methods used here is the viscous layer method. Using this method for riblet surfaces, the virtual origin of the mean streamwise velocity profile can be estimated. The results of this method for the riblet surfaces were compared with the results of the methods of Choi, and Bechert and Bartenwerfer. In this section none of the presented quantities are non-dimensionalized. In the next section the non-dimensionalized profiles of velocity, turbulent intensities and Reynolds stress correlations are presented, showing that the investigated riblet surfaces indeed possess the characteristics of drag reducing surfaces. Finally, some conclusions are presented.

2. Experiments

The measurements reported in this article were performed in the low-speed water channel in the Laboratory for Fluid Dynamics and Heat Transfer of the Technical University of Eindhoven in The Netherlands. This channel has a test section which is 7 m long, 30 cm wide and 30 cm high (Fig. 1). In this test section a false surface, with a length of 6 m, was mounted 123 mm above the bottom wall of the water channel. This false surface consists of several parts. The first part which is closest to the contraction of the water channel is a 2250 mm fixed glass surface. This glass surface is followed by several removable pieces with different lengths and different kinds of surface geometries. A tripping wire was placed 600 mm downstream of the nose of the glass surface to make sure that all measurements were performed in a fully developed, two dimensional, turbulent boundary layer. The mean stream velocity of this water channel can be adjusted between 0.0 and 400 mm/s, while the water temperature is kept at 20.00 ± 0.05°C.

The primary element of the measurement system is the two component HeNe Laser Doppler Anemometer (LDA). The computer system was updated during the last year, therefore the measurements were performed with two different control systems. At first the measurements were controlled by a PDP-11/23 computer, which was replaced by a PHYDAS system (physics data acquisition system) using a M68030 processor. This system was developed by the faculty of Technical Physics at the Eindhoven University of Technology. More detailed descriptions of the measurement system are given by Pulles (1988) and Kern (1985).

A two component LDA system uses three beams, two reference beams and one main beam, to obtain the velocities. In our case the two reference beams are parallel to the surface and the main beam makes an angle with the surface. The beams are further positioned in such a way that alternately a $u-v$ component and a $u+v$ component are measured. As the time interval between two successive samples is very small ($\leqslant 0.002$ s), the u and v can be obtained through interpolation between two successive

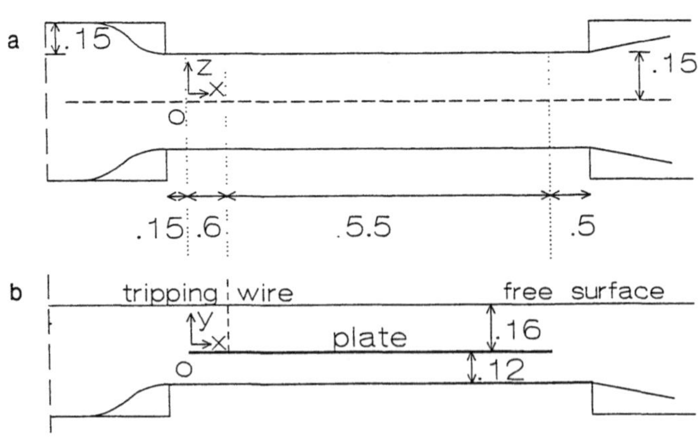

Fig. 1. Top (a) and side view (b) of the low-speed water channel.

samples. A detailed description of this setup is given by Schwarz-van Manen and Stouthart (1990).

This configuration of the laser beams can lead to two effects which appear when measurements are performed very close to the surface. First there is the backscatter of the main laser beam into the measurement volume due to reflection at the surface. A second effect is that the horizontal reference beams will eventually touch the surface, even before the measurement volume does. This is because the laser beams converge into a waist, which is at the position of the measurement volume, thereafter they diverge.

Special jigs were used to calibrate the vertical and spanwise position of the measurement volume with respect to the peak of the grooved surface (van Geloven 1992). The present LDA system has not only a vertical traversing mechanism but also a calibrated spanwise traversing system which permits the measurement volume to be located at different spanwise positions above the grooves. For safety reasons, the measurement volume could not get nearer than 100 μm to the surface. The measurement volume had a spanwise width of 0.5 mm and a vertical and streamwise width of 0.1 mm.

The measurements were performed over four different surfaces shown in Fig. 2: a smooth surface ($\mathscr{U}$); a triangular riblet surface ($\mathscr{R}$) with $h/s = 0.75$ and $\mathrm{tg}(\alpha) = 0.67$; a triangular riblet surface ($\mathscr{P}$) with $h/s = 0.9$ and $\mathrm{tg}(\alpha) = 0.5$; and a triangular surface ($\mathscr{S}$) which has the same dimensions as the $\mathscr{P}$-surface with every other triangle skipped, for which $h/s = 0.45$ and $\mathrm{tg}(\alpha) = 0.5$. Here α is half the apex angle of the ribs, s is the spacing between the centres of the grooves and h is their depth.

During the experiments the vertical profiles of the mean streamwise velocities, the turbulent intensities and the Reynolds stress correlations were measured. These measurements were used to obtain the friction velocity and the virtual origin. During a one day experiment it was possible to measure the above mentioned quantities at 40 different heights. This is due to the long averaging time ($\approx$ 10 min) caused by the much larger time scales in water experiments when compared with a wind tunnel. Typical values for the viscosity ν and the friction velocity u_* for the presented water channel measurements are $1 \cdot 10^{-6}$ m^2/s and 0.01 m/s, respectively, with a dimensionless time scale of 0.01 s. However, for a wind tunnel typical values for ν and u_* are $15 \cdot 10^{-6}$ m^2/s and 0.5 m/s, respectively, leading to a dimensionless time scale of $6 \cdot 10^{-5}$ s, which is roughly 2 orders of magnitude smaller than the time scale for the water channel measurements. Due to this long averaging time each profile measurement lasted 3 days. The first day was used to measure the outer layer (10–100 mm). The next day to measure the logarithmic region (1–10 mm) and a day was used to measure the very near wall layer (0.1–1 mm).

For all the experiments the water level above the mounted surface, at zero mean stream velocity, was kept at 158 mm. The measurements were performed at 404.5 ± 0.5 cm downstream of the nose of the mounted surface and 344.0 ± 0.5 cm downstream of the tripping wire. Further, experimental conditions are given in Table 1. The errors indicate here the maximum deviation from the average values.

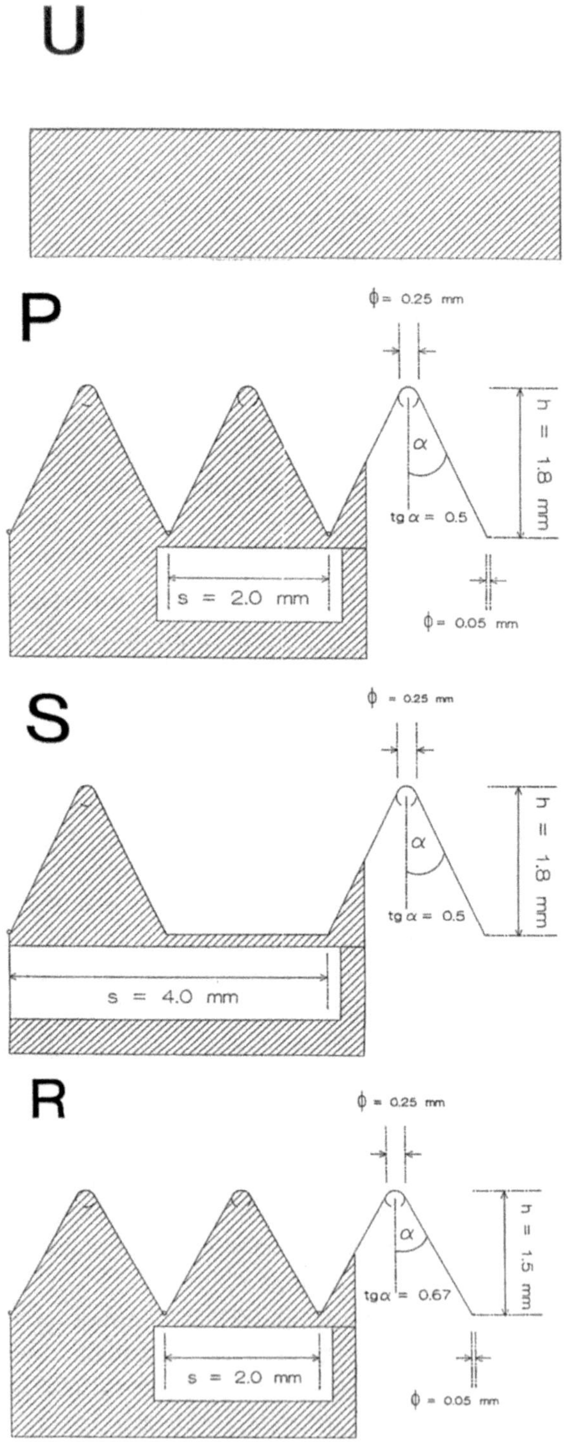

Fig. 2. Geometries of the investigated surfaces.

Table 1. Experimental conditions.

	$\mathscr{U}$	$\mathscr{R}$	$\mathscr{P}$	$\mathscr{S}$
U_0 [cm/s]	20.0 ± 0.3	20.0 ± 0.3	20.0 ± 0.3	20.0 ± 0.3
δ^* [mm]	14.6 ± 0.8	13.1 ± 1.5	14.5 ± 0.9	12.0 ± 1.3
θ [mm]	10.4 ± 0.5	9.4 ± 1.2	10.4 ± 0.8	8.6 ± 1.3
Shape factor	1.41 ± 0.04	1.40 ± 0.06	1.40 ± 0.04	1.40 ± 0.05
Re_x	$8 \cdot 10^5$	$8 \cdot 10^5$	$8 \cdot 10^5$	$8 \cdot 10^5$
Re_θ	2080	1880	2080	1720
T [°C]	20.0 ± 0.05	20.0 ± 0.05	20.0 ± 0.05	20.0 ± 0.05

3. Local friction velocity measurements

The friction velocity (u_*) is indicative of the amount of drag a flow experiences from a surface. When comparing the results of smooth and drag reducing riblet surfaces with each other it is important to know the u_* as accurately as possible, because the amount of drag reduction between a smooth and a drag reducing surface is about 5%.

In this section several methods will be discussed to obtain the friction velocity from the velocity profile. In the first subsection methods for smooth surfaces will be discussed while the second subsection deals with methods for grooved surfaces.

3.1. Methods for the smooth surface

When no drag balance is available there are various ways to estimate u_* from the velocity profile. The one most commonly used is the logarithmic friction law (Tennekes and Lumley, 1972). This law is valid in the logarithmic region of the flow and is defined as

$$\frac{\bar{U}}{u_*} = \frac{1}{\kappa} \ln\left(\frac{yu_*}{\nu}\right) + A. \tag{1}$$

In this expression, the kinematic viscosity ν is $1 \cdot 10^{-6}$ m^2/s at 20°C, $\bar{U}$ is the time average velocity and A is an empirical constant which, according to the literature, has a value between 4.9 and 5.45. Using the logarithmic region to estimate u_* the question will arise which value to use for the Von Kàrmàn constant κ ($0.38 \leqslant \kappa \leqslant 0.42$). Another problem is the exact estimation of the logarithmic region in a velocity profile. The standard region mentioned in the literature is

$$30 \leqslant \frac{yu_*}{\nu} \leqslant 800, \tag{2}$$

but it is mainly determined by the Reynolds number. Therefore, the proper range where the logarithmic region can be found has to be estimated from the velocity profiles.

Bradshaw, Launder and Lumley (1991) made the following remark about using the logarithmic region to estimate the friction velocity:

> ... One remarkable cause of confusion is the wide range of constants in the logarithmic "law of the wall" that are assumed, or predicted, in different methods. According to first order arguments, which are used implicitly in all the calculation methods, these constants should be universal. The actual range – ignoring a few outlying methods – implies a 10% "uncertainty" in skin friction on a flat plate in a low-speed flow, the simplest of all test cases. We are trying to achieve a consensus! ...

A second method which can be used is the Clauser method (1956). This method uses the complete velocity profile of a turbulent boundary layer to estimate the u_*. Clauser made use of the following relations:

$$u_* = U_0 \sqrt{\frac{C_f}{2}}, \tag{3}$$

$$\left(\frac{C_f}{2}\right)^{-1/2} = \frac{1}{\kappa} \ln Re_{\delta*} + B, \tag{4}$$

with

$$Re_{\delta*} = U_0 \delta^*/\nu. \tag{5}$$

The displacement thickness δ^* is defined by

$$\delta^* = \int_0^\infty (1 - \bar{U}/U_0)\, \partial y. \tag{6}$$

In these equations C_f is the friction coefficient, U_0 is the mean free stream velocity and B an empirical constant and equals, according to the literature, 4.9.

The advantage of this method compared with the previous is that it uses the whole velocity profile and not a part of it, which has to be estimated by eye. However one problem still remains; again an assumption has to be made for the Von Kàrmàn constant κ.

To avoid problems with κ and the empirical constants, the velocity profile of the viscous layer was used to obtain the friction velocity. The following expression is valid for this layer in the region $y^+ \leqslant 5$

$$\tau_y = \nu \frac{\partial \bar{U}}{\partial y} = \frac{\tau_0}{\rho} + y \frac{\partial P}{\partial x}. \tag{7}$$

Since, $\partial P/\partial x < 1.8$ Pa/m and $y^+ \leqslant 5$ the following relation is valid for the friction velocity u_*,

$$u_* = (\tau_0/\rho)^{1/2}. \tag{8}$$

In these expressions τ_0 is the wall shear stress and ρ is the density. Combining equations (7) and (8) the following relation is obtained

$$\bar{U} = \frac{u_*^2}{\nu}\, y. \tag{9}$$

Although this method has the disadvantage of estimating the range of the viscous layer by eye, the error will be much smaller than the approximately 5% to 10% error of the other two methods.

3.2. Methods for the grooved surfaces

The methods discussed in the previous section are only valid for the smooth surface, because the positions of $y = y^+ = 0$ are exactly known. Problems arise when these methods have to be used for grooved or rough surfaces, where the top of the roughness is normally calibrated as $y = 0$. The virtual origin of the mean streamwise velocity profile lies somewhere between the top and the valley of a groove. It depends on the spanwise position of the measurement volume, for instance when the measurements are performed exactly above the top of a groove the friction velocity should have a maximum value and the virtual origin should be at $y = 0$. When on the other hand, the measurements are performed over the valley of a groove, the friction velocity should have a minimum and the difference between $y = 0$ and the virtual origin should reach a maximum value. Measurements performed over the slope of a groove should have values lying between those for the measurements above the top and valley. The viscous layer method is the easiest method which can be made applicable for grooved surfaces. Accounting for the virtual origin equation (9) will be modified as

$$\bar{U} = \frac{u_*^2 y_v}{\nu} = \frac{u_*^2(z)}{\nu}\,(y + \Delta y_v(z)), \tag{10}$$

where $y_v(z)$ is the transformed coordinate and Δy_v the displacement of the origin. From the viscous layer velocity profiles the gradient $u_*^2(z)/\nu$ and the constant $u_*^2(z)\Delta y_v(z)/\nu$ can be estimated, from which the *local* values for the friction velocity $u_*(z)$ and the displacement of the origin $\Delta y_v(z)$ can be obtained.

The substitution of $\Delta y_v(z)$ can also be applied to the log-law method, it was previously done by Perry and Joubert (1963), Hooshmand et al. (1983) and Clauser

(1954). They showed that a logarithmic law could describe the mean streamwise velocity in the near wall region of a rough wall pipe, when it was formulated as follows:

$$\frac{\bar{U}}{u_*(z)} = \frac{1}{\kappa} \ln\left(\frac{yu_*(z)}{\nu} - \frac{\Delta\bar{U}}{u_*(z)} \frac{\Delta y_v(z)u_*}{\nu}\right) + A. \tag{11}$$

Due to the fact that this equation contains two unknowns it cannot be solved.

Another method that can be used to obtain the friction velocity, u_*, and the transformed coordinate, y_v, for the grooved surfaces is the method described by Choi (1989), which is a variant of the Furuya and Fujita (1966) method. For the Furuya and Fujita method the u_* for the grooved surfaces must be known, to estimate the virtual origin; Choi changed this method slightly, by plotting $(U_0 - \bar{U})/u_*$ vs. $yu_*/\delta^* U_0$ for the grooved surface in the same graph as for the smooth surface. As a start of this iterative method an assumption is made for the u_* and the y_v of the grooved surface. In the region between $0.002 \leqslant yu_*/\delta^* U_0 \leqslant 0.15$ the profiles for the grooved and the smooth surface should be identical. Therefore, a fourth-order polynomial is fitted through both profiles to check this. In order to obtain the best match between these two polynomials the two parameters u_* and y_v for the grooved surface measurements are modified at every iteration, thus also estimating δ^* before every iteration. Within a few iterations the values for u_* and y_v for the grooved surface can be found. In Fig. 3 a graph is shown of the Choi method after the best values of u_* and y_v had been found for the profile above a grooved surface.

The last verification used is the theoretical model of Bechert and Bartenwerfer (1989). They developed a model to estimate isovelocity lines of the mean stream velocity above grooved surfaces, by starting with the Navier–Stokes equation and neglecting the time-dependent terms as only the mean flow is considered. The pressure term was neglected also, because a flat plate boundary layer flow was assumed. Furthermore, the convective terms can be neglected when the assumption is made that the riblets are completely immersed in the viscous layer. The cross section of the riblets does not vary in the streamwise direction. Therefore, they obtained a Laplace

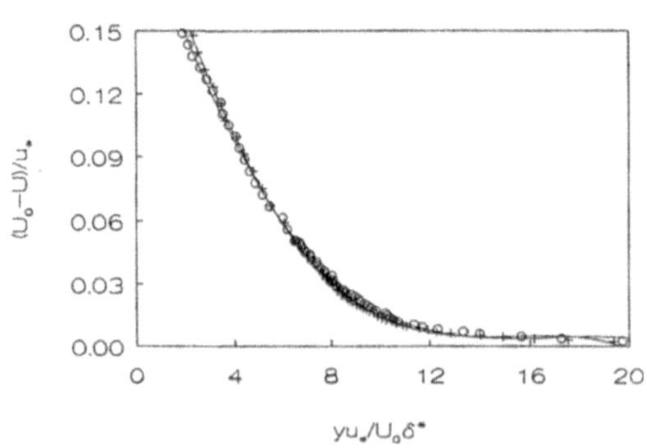

Fig. 3. An example of the Choi method (1989). Smooth surface +, grooved surface ○.

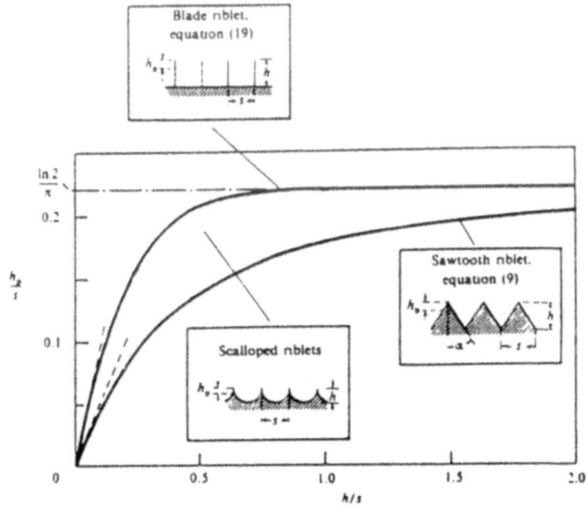

Fig. 4. Protrusion height h_p as a function of geometrical height h for various types of riblets.

equation for the mean streamwise velocity $\bar{U}$. After using conformal mapping they could calculate the protrusion height for different surface geometries, as plotted in the graphs in Figs 4 and 5. The protrusion height is a quantity which indicates how far the top of the grooves penetrated into the viscous layer. In Section 4.2 a comparison between the protrusion height and the virtual origin will be made.

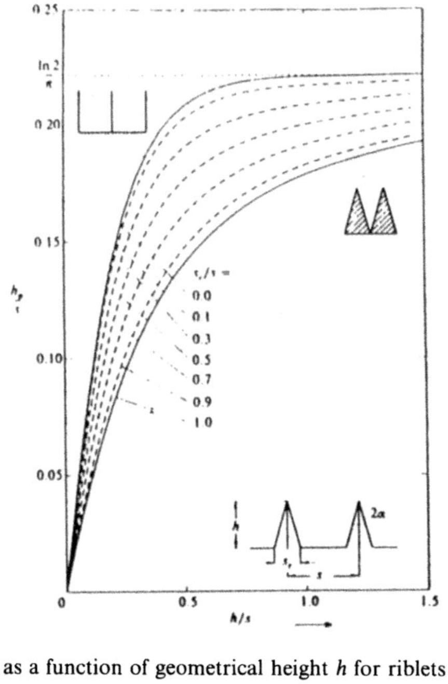

Fig. 5. Protrusion height h_p as a function of geometrical height h for riblets with trapezoidal valleys. The riblet foot breadth s_r is kept constant for each curve.

4. Results

The results of the experiments will be discussed in three phases. In the first subsection only the results for the smooth surface will be discussed and the constant κ will be estimated by using both the Clauser method and the law of the wall method with the friction velocity obtained from the viscous region. After this subsection the results for the grooved surfaces will be discussed. Here the effect of the spanwise position of the measurement volume will be shown. We then indicate the conditions under which the measurements of the very near wall region can or cannot be used to estimate the friction velocity and the protrusion height. Furthermore a comparison will be given between the Choi method, the method of Bechert and Bartenwerfer and the viscous region. In the last subsection the effect of using different scaling parameters will be shown.

4.1. Results for the smooth surface

In this section only the results for the smooth surface will be discussed and the constant κ will be estimated by using both the Clauser method and the law of the wall method, using the friction velocity obtained from the viscous region.

The measurement volume of the LDA system could be positioned as low as 0.32 ± 0.02 mm above the smooth surface; below this position it was impossible to obtain any signal. The viscous layer above the smooth surface was approximately 0.7 mm thick; the velocity profile of this layer was measured twice (Fig. 6). Very close to the surface the wall effect can be noticed (see Section 2). It looks very similar to the wall effect which arises in a wind tunnel when hot-wires are used to measure the velocity (Bessem 1991, Stouthart 1991 and Van Sas 1982). The friction velocity, u_*, was determined from the velocity profile between $0.4 \leqslant y \leqslant 0.7$ mm; in this region about 20 measurements were performed. This region was chosen because below

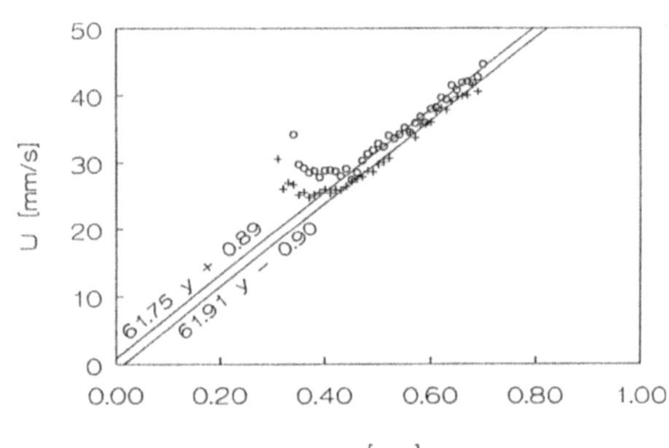

Fig. 6. Velocity profile of the viscous sublayer above a smooth surface. $U_0 = 20.0$ cm/s. Each symbol represents an independent measurement.

0.4 mm the wall effect is observed and above 0.7 mm the transition of the viscous layer into the buffer layer starts. The latter boundary was established by iteration; every time after the u_* was obtained, it was verified whether the used region stayed below 5 viscous length units. The average value for u_* for the two plots (Fig. 6) was 7.83 ± 0.06 mm/s.

Figure 7 shows the total profile of the boundary layer over a smooth surface, each symbol representing the part of the velocity profile measured in one day. Figures 6 and 7 show that the reproducibility of the measurements is good; for the estimation of u_* a reproducibility of 1% was achieved. From Fig. 7 it was estimated that the logarithmic region lies between $30y^+$ and $100y^+$. The Von Kàrmàn constant κ was determined from this region using the log-law (equation (1)) and the friction velocity which was obtained from the viscous layer. The average Von Kàrmàn constant for these experiments at a mean stream velocity of $U_0 = 20$ cm/s was $\kappa = 0.38 \pm 0.01$.

The displacement thickness, δ^*, was estimated from the complete velocity profile, using equation (6). The results are given in Table 1. This value was used in the Clauser method together with the u_* from the viscous layer to obtain the Von Kàrmàn constant, resulting in an average value for κ of 0.39 ± 0.01.

If the normal value 0.4 were used for κ to estimate the friction velocity, either with the Clauser method or the log-law method, an uncertainty between 2% and 5% would have been introduced. Therefore, when no drag balance is available and there is a possibility to measure the viscous layer velocity profile, it is strongly recommended to use this layer to estimate the friction velocity.

4.2. Results for the riblet surfaces

The measurement volume of the LDA system could be as close as 0.12 mm to the $\mathscr{S}$- and $\mathscr{R}$-surfaces and about 0.20 mm to the $\mathscr{P}$-surface. This was due to the fact that the back scattering of the main laser beam was a smaller problem above the grooved

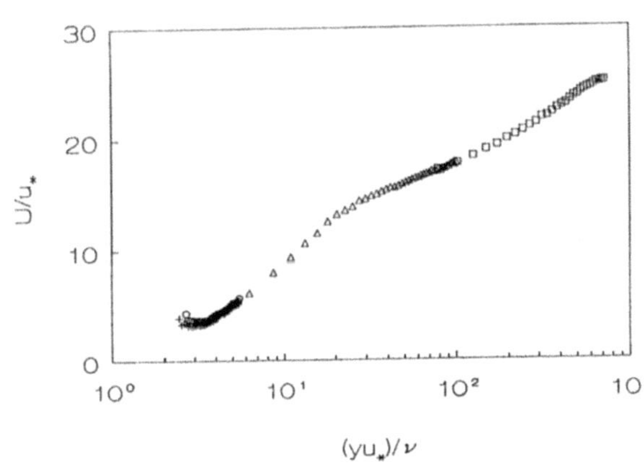

Fig. 7. A semilogarithmic plot of the mean stream velocity profile above a smooth surface. Each symbol represents measurements on a single day.

surface than above the smooth surface. As mentioned in Section 2 no measurements could be performed below 100 μm.

Figure 8 is a semi-logarithmic graph of the mean streamwise velocity for the four investigated surfaces, Fig. 9 shows a linear graph of the very near wall region. In both graphs, close to the surface the wall effect can be noticed for the $\mathscr{P}$- and $\mathscr{U}$-surfaces. One of the reasons that this effect is not seen for the other surfaces is that the wall effect is below 100 μm. The reason that the measurement volume could not come lower than 0.2 mm above the $\mathscr{P}$-surface may lie in the fact that this surface was slightly concave in the spanwise direction. The reference beams were disturbed by the edges of this surface.

The effects of the grooves on the velocity profiles can be noticed in the near wall region by their upward shift compared to the same profile for the smooth surface. Especially Fig. 9 shows very nicely that the origin of the profiles for the grooved surfaces lies below the top of the grooves; $y = 0$ was calibrated at the top of the grooves (Section 2). This effect is not only characteristic for drag reducing surfaces, but is characteristic for rough surfaces in general (Grass, 1971).

The velocity profile of the $\mathscr{S}$-surface lies much above the profiles of the $\mathscr{P}$- and $\mathscr{R}$-surfaces, which suggests that the virtual origin for the $\mathscr{S}$-surface lies lower than for the other two surfaces. The maximum difference between the virtual origin and the calibrated origin is different for each rough surface. Only when the position of the virtual origin of the velocity profiles is found, can proper comparisons between the smooth and the grooved surfaces be made.

Measurements with exactly known spanwise positions were carried out only for the $\mathscr{P}$- and $\mathscr{S}$-surfaces. Figure 10 shows the velocity profiles of the very near wall region above a riblet top, a side and a bottom of a groove of the $\mathscr{P}$-surface, each symbol referring to a different spanwise position (Fig. 10b) of the measurement volume. In Fig. 10a the effect of the spanwise position above the grooves can be noticed by the

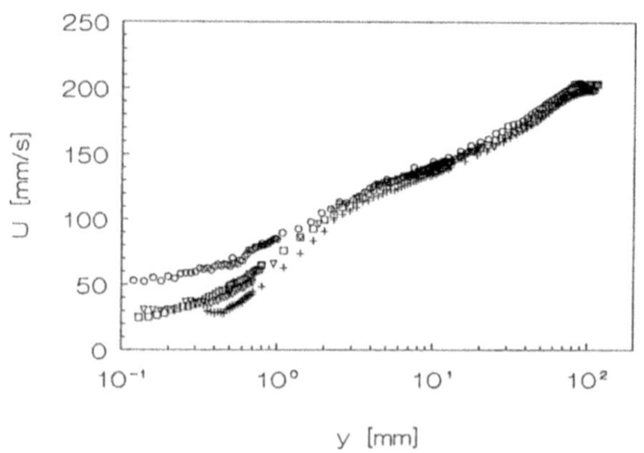

Fig. 8. A semi-logarithmic plot of the mean stream velocity profiles, with $\mathscr{U}$ +, $\mathscr{P}\nabla$, $\mathscr{R}\square$ and $\mathscr{S}\bigcirc$ and $y = 0$ at the top of the riblets.

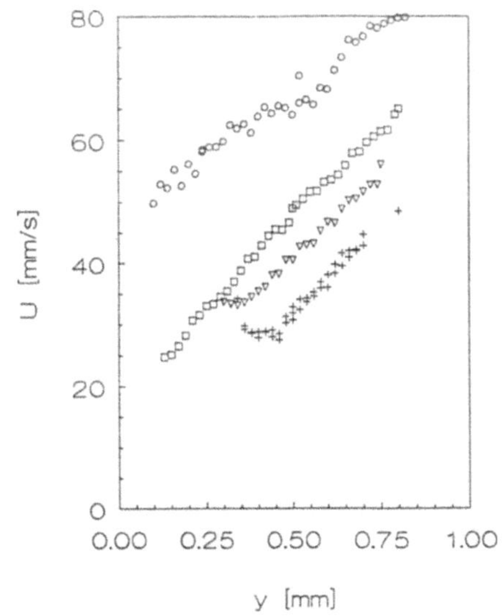

Fig. 9. The mean streamwise velocity profiles in the very near wall region, with 𝒰 +, 𝒫∇, ℛ□ and 𝒮○ and $y = 0$ at the top of the riblets.

fact that the highest profile is the one measured above the bottom of a groove and the lowest profile the one measured over the top of a riblet. This result can be explained with Fig. 11. Above a certain height, ($y^+ \approx 13$), above the grooves the mean streamwise velocity profiles are identical, while the virtual origins of the velocity profiles depend on the spanwise position of the measurement volume. Therefore, the

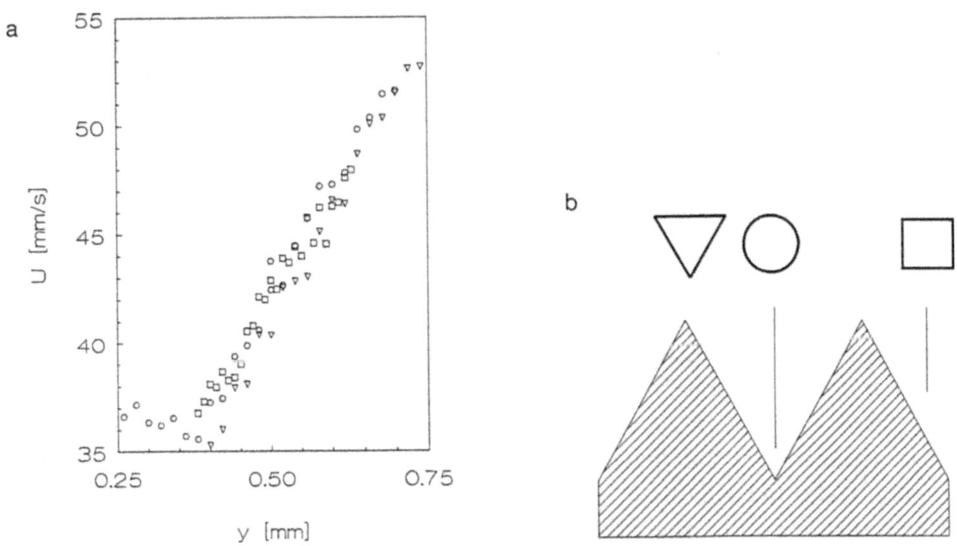

Fig. 10. The velocity profiles above (a) the 𝒫-surface. Each symbol represents a different spanwise position (b) and $y = 0$ is at the top of the riblets.

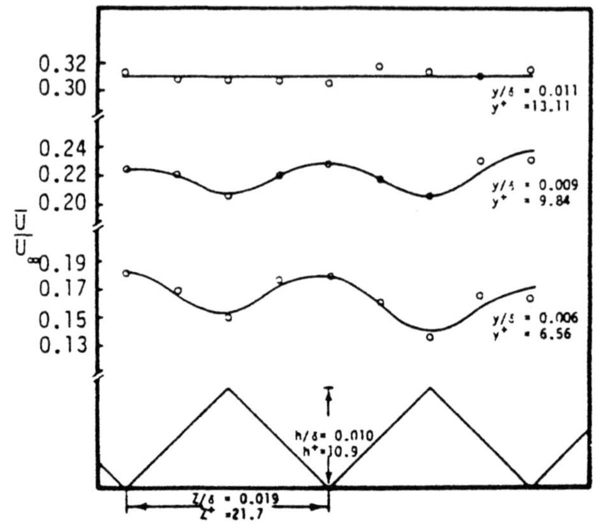

Fig. 11. Variation of mean velocity with spanwise location at several heights above a riblet surface.

gradient of the mean streamwise velocity profile above the top of a riblet is steeper than for the profile over the bottom of a groove. According to this discussion, one would expect after calibrating $y = 0$ at the top of a riblet, that the velocity would be zero at $y = 0$ above the top of a riblet and has a certain value above the bottom of a groove. However, after extrapolating the velocity measured the top of a riblet it does not become zero at $y = 0$, this is because the measurement volume has a spanwise width of 0.5 mm. This implies that the measured velocity is an average velocity over a spanwise distance of 0.5 mm and therefore cannot be zero above a top of a groove. For instance, for the $\mathscr{P}$-surface the measurement volume will always cover 25% of the groove. Another result of the above discussion is that the measurements above the side of the groove are expected to lie in between the results for the measurements above the top and the bottom of a groove, as shown in Fig. 10a. These results agree with the experiments of Hooshmand et al. (1983) shown in Fig. 11. They measured the mean streamwise velocity component in and out of the grooves at several heights and spanwise positions.

So far, only a qualitative comparison has been made between the experiments. Before it is possible to make a quantitative comparison between the experiments, the position of the viscous region has to be determined. For the first order the characteristics of the viscous region can be described by: (i) a linear profile for the mean stream velocity; (ii) low values for the Reynolds stress compared to the viscous stress; and (iii) this region lies below $y^+ = 5$.

Ignoring the region where the wall effects impair the measurements, Fig. 9 shows that except for the $\mathscr{S}$-surface all other surfaces exhibit a region with a linear profile. In spite of this fact, the viscous region for all four turbulent boundary layers was estimated, and both the friction velocity, u_*, and the displacement of the origin, Δy_v, were determined with the viscous layer method (see Table 2).

Table 2 summarizes the results obtained for the friction velocity, the displacement of the origins and the protrusion height using the three different methods. Both the results for the viscous method and the Choi method coincide with one another, but the Choi method has some disadvantages compared to the viscous method. First of all the computation is very time consuming, secondly the method is more sensitive to changes in u_* rather than Δy_v and therefore the latter is not determined very accurately. Furthermore, it can be noticed that results of the viscous method are the most consistent. Comparing the displacement of the origin above the valleys of the grooves for both the viscous and Choi method one notices that they are almost equal to the protrusion height predicted by the theory of Bechert and Bartenwerfer (obtained from Figs 4 and 5). However, here the difference is quite large for the $\mathscr{S}$-surface. This is probably due to the fact that the turbulent intensity profiles and the Reynolds stress correlation (Fig. 12) showed a significant difference for the $\mathscr{S}$-surface compared to the other surfaces. Although, the maximum turbulent intensity (u_{rms}) for this surface is the lowest of all the experiments, it shows significantly higher values for the turbulence intensities and the Reynolds stress ($\overline{u'v'} \simeq 1/3u_*^2$) below 0.7 mm. Secondly, the viscous layer was difficult to establish. It might as well lie within the riblets. Therefore, the viscous method results for the $\mathscr{S}$-surface should be looked upon with some caution. The viscous method results for the other surfaces show no variation. In Table 3 the percentage differences between the methods are shown.

The estimation of the viscous layer above the riblet surfaces was done in the same manner as for the smooth surface (Section 4.1). From Table 4 it can be seen that these grooved surfaces introduce an increase in thickness of the viscous layer. There is one remarkable thing however. In Table 2 the respective displacement of the origins of the measurements are given and if these heights are subtracted from the viscous layer thickness (Table 4), the remaining heights for these surfaces become equal to the thickness of the viscous layer above the smooth surface. This suggests that the peaks of these drag reducing riblet surfaces do not penetrate beyond the viscous layer. This

Table 2. u_{*l} and h_{pl} obtained by several methods and for different spanwise positions.

Surface		Viscous layer		Choi method		B & B method
		u_* [mm/s]	Δy_v [mm]	u_* [mm/s]	Δy_v [mm]	h_p [mm]
$\mathscr{U}$		7.83 ± 0.06	0			
$\mathscr{P}$	valley	7.07 ± 0.07	0.35 ± 0.02	6.9 ± 0.1	0.3 ± 0.1	0.35
	slope	7.11 ± 0.07	0.34 ± 0.01			
	top	7.27 ± 0.09	0.26 ± 0.02	7.3 ± 0.1	0.3 ± 0.1	
$\mathscr{S}$	valley	7.55 ± 0.10	0.57 ± 0.05	7.5 ± 0.1	0.6 ± 0.1	0.68
	top	7.59 ± 0.06	0.50 ± 0.03	7.5 ± 0.1	0.6 ± 0.1	
$\mathscr{R}$		7.69 ± 0.08	0.31 ± 0.02	7.4 ± 0.1	0.3 ± 0.1	0.31

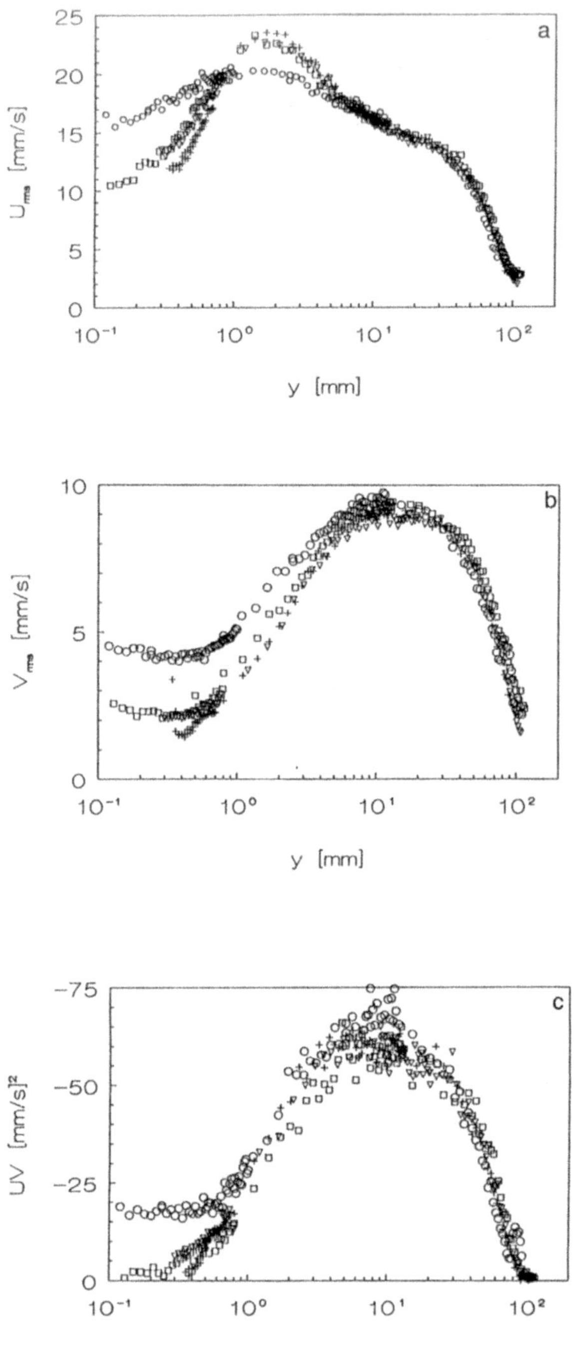

Fig. 12. The turbulent intensity profiles (a, b) and the Reynolds stress correlation (c). $\mathscr{U}$ +, $\mathscr{P}\nabla$, $\mathscr{R}\square$ and $\mathscr{S}\bigcirc$.

Table 3. Difference between the results of the three methods.

Surface	$\left\|\dfrac{u_{*\,\text{Choi}} - u_{*\,\text{vis}}}{u_{*\,\text{vis}}}\right\|$	$\left\|\dfrac{\Delta y_{v\,\text{Choi}} - \Delta y_{v\,\text{vis}}}{\Delta y_{v\,\text{vis}}}\right\|$	$\left\|\dfrac{h_p - \Delta y_{v\,\text{vis}}}{\Delta y_{v\,\text{vis}}}\right\|$	$\left\|\dfrac{\Delta y_{v\,\text{Choi}} - h_p}{h_p}\right\|$
$\mathscr{P}$ valley	0.024	0.143	0.0	0.143
top	0.004	0.154		
$\mathscr{S}$ valley	0.007	0.053	0.193	0.118
top	0.012	0.200		
$\mathscr{R}$	0.038	0.032	0.0	0.032

Table 4. The average thickness of the viscous layer estimated from the velocity profiles given in mm and dimensionless length scales.

Surface	Thickness viscous layer [mm]	l^+
$\mathscr{U}$	0.6	4.7
$\mathscr{P}$ valley	0.95	6.7
slope	0.89	6.3
top	0.86	6.3
$\mathscr{R}$	0.93	7.2
$\mathscr{S}$ valley	1.17	9
top	1.04	8

was essentially one of the basic assumptions of the work of Bechert and Bartenwerfer (1989).

4.3. Dimensionless profiles

For both the displacements of the origin and the friction velocity of the grooved surface the spanwise average values of the viscous layer method were used, except for the $\mathscr{S}$-surface where the results of the Choi method were used, because the viscous method was not completely reliable for this surface. Bandyopadhyay (1986) also advised to use the spanwise average value of these quantities; because above $\sim 13y^+$ there is no difference in the spanwise variation of the mean streamwise velocity. After the origin displacement correction (Table 5) was applied to the measurements the displacement thickness δ^* and the momentum thickness θ (Table 1) were estimated. The profiles were non-dimensionalized with different characteristic variables. Figure 13 shows two semi-logarithmic graphs of the mean streamwise velocity profiles. In Fig. 13a the inner velocity scale $u_i^+ = u_*$ and inner length scale $l_i^+ = \nu/u_*$ were used, while for the other graph the velocity and length scale were obtained using

Table 5. The average values of the friction velocity and the displacement of the origin.

Surface	u_* [mm/s]	Δy_v [mm]
$\mathscr{U}$	7.83	
$\mathscr{P}$	7.17	0.31
$\mathscr{S}$	7.5	0.6
$\mathscr{R}$	7.69	0.31

the free stream velocity U_0 and the displacement thickness δ^*. This results in our outer length scale $l_0^+ = \delta^*$ and the outer velocity scale $u_0^+ = U_0$.

When these figures are compared with Fig. 8 the transformation of origin is clearly noticeable. In the viscous region of the velocity profile the difference between the profiles has become much smaller, but for the $\mathscr{S}$-surface still a small upward shift can be seen. The effect of using the friction velocity to non-dimensionalize the graphs is clearly noticeable in the logarithmic and outer region. If the friction velocity of the

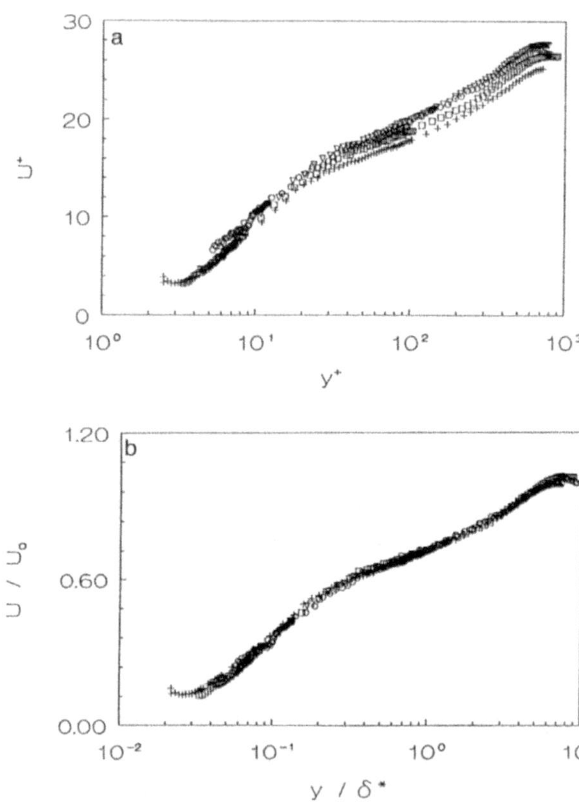

Fig. 13. The mean streamwise velocity: (a) using the inner scales $l_i^+ = \nu/u_*$ and $t_i^+ = \nu/u_*^2$; (b) using the outer scales $l_0^+ = \delta^*$ and $t_0^+ = \delta^*/U_0$. With $\mathscr{U}$ +, $\mathscr{P}\nabla$, $\mathscr{S}\bigcirc$, $\mathscr{R}\square$.

grooved surfaces is smaller than for the smooth surface (drag reduction), the dimensionless profiles will shift upwards compared to the profile of the smooth surface. In the case of a larger friction velocity (drag increase) the profiles will shift downwards (Grass, 1971).

Comparing Figs 13a and 13b it must be concluded that using the displacement thickness as a length scale gives the best results. Regarding the turbulence intensity profiles and the Reynolds stress correlation (Figs 14 and 15), the difference between the scaling can only be found for the turbulent intensities u_{rms} and v_{rms}. When the outer scales (δ^* and U_0) are used the best match between the profiles is obtained. But still a difference can be noticed between the surfaces for the maximum u_{rms}. The results of Hooshmand (1983), Pulles (1988) and Choi (1989) showed that for drag reducing surfaces the maximum $u_{rms}/\bar{U}$ was amaller than for the smooth surface. Figure 15 suggests, therefore, that the $\mathscr{S}$-surface should have a larger drag reducing effect than the $\mathscr{R}$- and $\mathscr{P}$-surface. This is in contradiction with the friction velocities that were found. Scaling with the inner scales the profiles are shifted in such a way that it is very difficult to draw any conclusions.

5. Discussion

It can be stated that the experiments were reproducible within 1%. When no drag balance is available the measurements presented here strongly recommend the use, if possible, of the viscous region to estimate both the friction velocity u_* and the displacement of the origin Δy_v. This method has the advantage of being independent of the surface geometry, therefore no reference measurements are needed, as with for instance the Choi method. If the viscous region cannot be used the Choi method is the next best to use.

The comparison of the viscous layer method with the calculations of the method used by Bechert and Bartenwerfer for the $\mathscr{P}$- and $\mathscr{R}$-surface are extremely good; the protrusion height is equal to the displacement of the origin above the bottom of a groove.

That δ^* turned out to be a very good scaling parameter, this is probably due to the fact that it is influenced by the grooves.

The measurements presented here are all local measurements. It is therefore, difficult to draw strong conclusions about the drag reducing behavior of the different surfaces. With this reservation in mind, the measurements suggest that the $\mathscr{P}$-surface provides the maximum drag reduction. This surface has the largest h/s value and the smallest top angle $\tan\alpha$. Comparing the present results and the work of Bechert and Bartenwerfer (1989), it can be concluded that scalloped grooves and vertical blades in the streamwise direction will show an even larger drag reduction. The groove spacing s and the height of the grooves h must of course have the proper dimensions, their ratio (h/s) must be close to 1 and the top angle of the grooves ($\tan\alpha$) as small as possible.

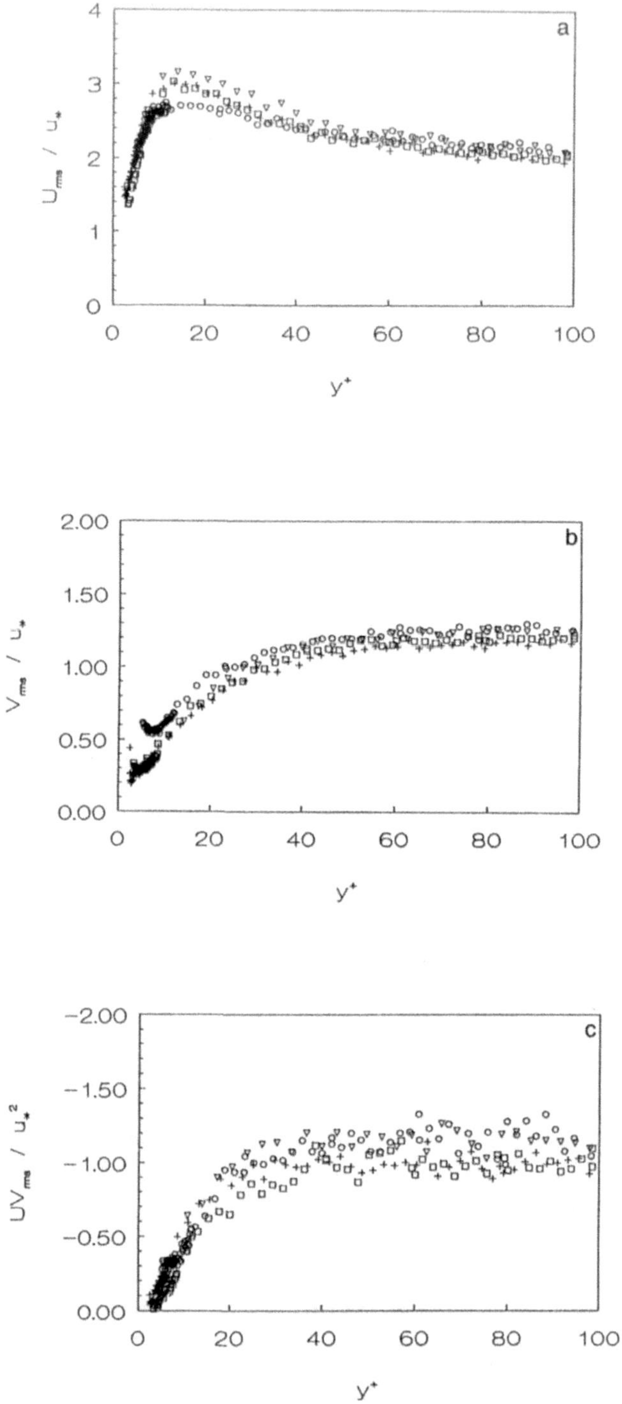

Fig. 14. Turbulence intensities and Reynolds stress correlation non-dimensionalized with the inner scales. $\mathscr{U}$ +, $\mathscr{P}$ ∇, $\mathscr{S}$ ○, $\mathscr{R}$ □.

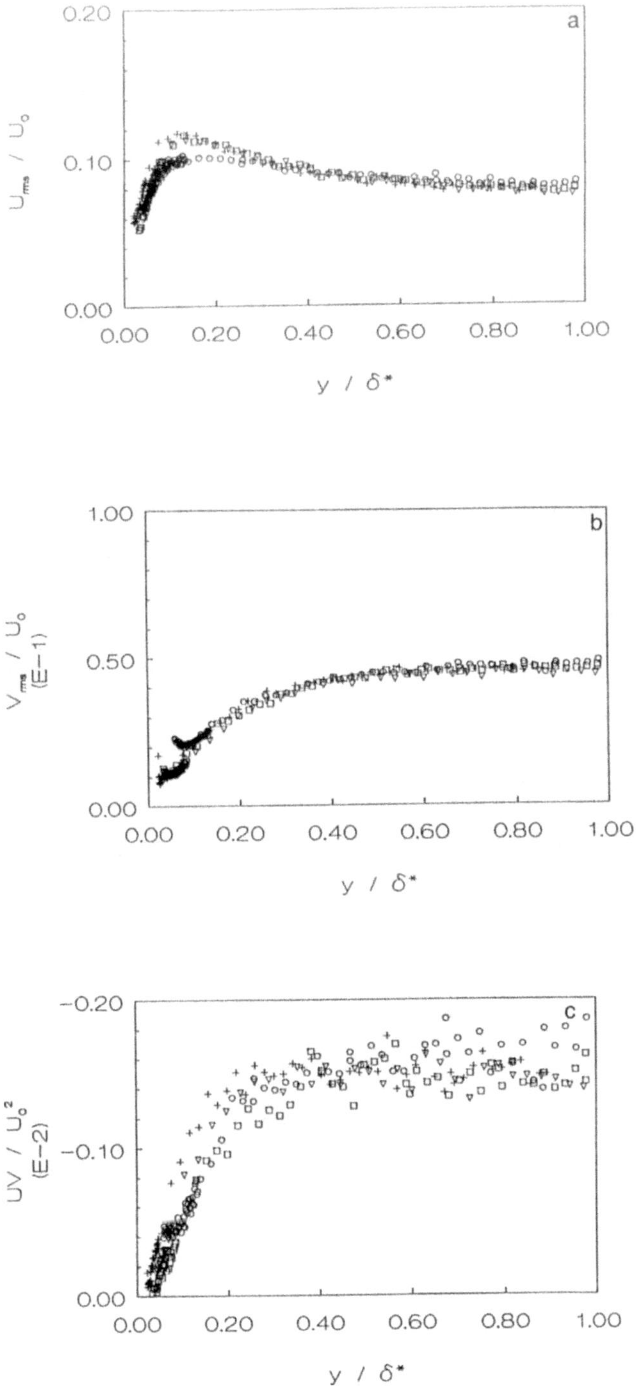

Fig. 15. Turbulence intensities and Reynolds stress correlation non-dimensionalized with the outer scales. $\mathscr{U}$ +, $\mathscr{P}$ ∇, $\mathscr{S}$ ○, $\mathscr{R}$ □.

6. Acknowledgement

The research in this article was supported by The Netherlands Foundation for Technical Sciences (STW) and the Foundation for Fundamental Research on Matter (FOM).

7. References

Bandyopadhyay, P.R., Review – Mean flow in turbulent boundary layers disturbed to alter skin friction. *J. Fluid Eng.* 108 (1986) 127.

Bessem, J.M., personal communication (1991).

Bradshaw, P., Launder, B.E. and Lumley, J.L. Collaborative testing of turbulence models. *J. Fluids Eng.* 113 (1991) 3.

Bechert, D.W. and Bartenwerfer, M., The viscous flow on surfaces with longitudinal riblets. *J. Fluid Mech.* 206 (1989) 105.

Choi, K.S., Near-wall structure of turbulent boundary layers over smooth and riblet surfaces. *J. Fluid Mech.* 208 (1989) 417.

Clauser, E.H., Turbulent boundary layers in adverse pressure gradients. *J. Aeronautical Sciences* 21 (1954) 91.

Clauser, E.H., Turbulent boundary layer. *Advances in Applied Mechanics* 4 (1956) 1.

Hooshmand, A., Youngs, R.A., Wallace, J.M. and Balint, J.L., An experimental study of changes in the structure of a turbulent boundary layer due to surface geometry changes. *AIAA* paper no. 83-0230 (1983).

Geloven, J. van, u_* en h_p bepaling door middel van verschillende methoden boven vlakke en gegroefde wanden. Internal report no. R-1153-S, Eindhoven University of Technology, The Netherlands (1992).

Grass, A.J., Structural features of turbulent flow over smooth and rough boundaries. *J. Fluid Mech.* 50 (1971) 233.

Furuga, M. and Fujita, H., Turbulent boundary layer over rough gauze surfaces. *Trans. Japan Soc. Mech. Engrs.* 32 (1966) 725.

Kern, J.H.A., Meting van de turbulente schuifspanning met een laser doppler anemometer in de grenslaag boven een vlakke en een ruwe wand. Internal report no. R-710-A, Eindhoven University of Technology, The Netherlands (1985).

Perry, A.E. and Joubert, P.N., Rough-wall boundary layers in adverse pressure gradients. *J. Fluid Mech.* 17 (1963) 193.

Pulles, C.J.A., Drag reduction of turbulent boundary layers by means of grooved surfaces. Ph.D. thesis. Eindhoven University of Technology, The Netherlands (1988).

Sas, L.M.L.A. van, Wandinvloed op snelheidsmetingen met behulp van een hetedraad-anemometer in een turbulente grenslaag. Internal Report no. R-520-S, Eindhoven University of Technology, The Netherlands (1982).

Schwarz-van Manen, A.D. and Stouthart, J.C., Beschrijving van de LDA opstelling van het waterkanaal. Internal Report no. R-1061-D, Eindhoven University of Technology, The Netherlands (1990).

Tennekes, H. and Lumley, J.L., *A First Course in Turbulence*. Cambridge (Mass.): The MIT Press (1972). ISBN 0-262-200-19-8.

Applied Scientific Research **50**: 255–266, 1993.
K. Krishna Prasad (ed.), *Further Developments in Turbulence Management.*

Viscous sublayer analysis of riblets and wire arrays

PAOLO LUCHINI
Istituto di Gasdinamica, Facoltà di Ingegneria, Università di Napoli, Italy

1. Introduction

Ever since it was observed [1] that the experimentally effective spacing of riblets for turbulent drag reduction is of the same order of magnitude as the thickness of the viscous sublayer of the turbulent stream, it has been clear that the viscous sublayer plays a dominant role in the mechanics of this effect. Within the viscous sublayer we are allowed to neglect nonlinearities and thus to concentrate on the Stokes, low Reynolds number, flow of a viscous fluid along a corrugated surface. Bechert and Bartenwerfer [2] analysed longitudinal mean flow in this approximation and obtained a characterization of the surface in terms of the parameter "protrusion height", namely the distance from the riblet tips to the virtual plane surface that the velocity profile appears to originate from. Bechert et al. [3] studied transverse flow across the riblets experimentally and by means of an analogical simulation, and measured a protrusion height for this flow as well.

In recent papers [4, 5, 6] some coworkers and I obtained analytical and numerical solutions for both the longitudinal and transverse flow problems, and proposed the difference between the longitudinal and transverse protrusion heights as a parameter characterizing the damping action of riblets upon streamwise vortices. The analytical results enabled us to prove that this difference is positive for shallow grooves of whatever shape and is again positive for infinitely deep grooves. The numerical results obtained for grooved surfaces of cosinusoidal, triangular and parabolic profile led us to conjecture that this parameter is always positive.

In Ref. [7], and in their paper included in this same volume, Bartenwerfer and Bechert reconsidered Kramer's early proposal [8] of using regular arrays of longitudinal wires suspended above a plane wall (see Fig. 1) as alternative drag-reduction devices with an operation mechanism similar to that of riblets. They calculated, by a conformal-mapping technique, the mean flow that takes place in the direction parallel to the wires when these are embedded inside the viscous sublayer of a turbulent stream. More recently Bechert et al. [9] have also performed an experiment on such a wire array, demonstrating some drag reduction.

At the suggestion of D. W. Bechert we, too, began to study arrays of longitudinal cylindrical wires suspended above a plane wall. In this paper we shall present approximate analytical solutions for both longitudinal and transverse Stokes flow about such a structure.

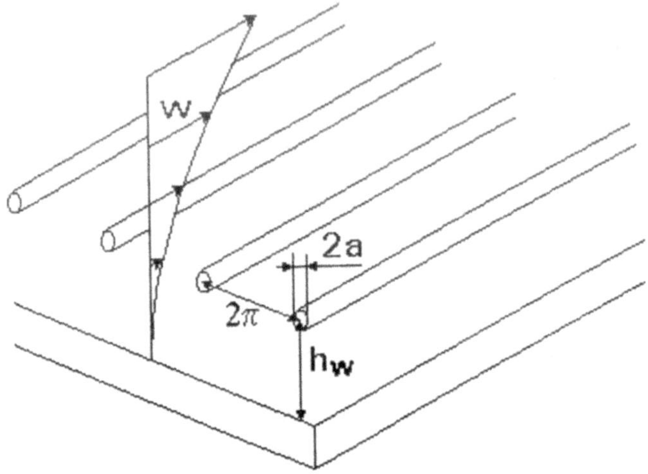

Fig. 1. Schematic representation of a wire-array drag-reduction device.

2. Role of the viscous sublayer in the operation of riblets and wire arrays

It is now a consolidated experimental fact that the best drag-reduction performance of riblets is obtained when their repetition period s^+ in viscous units is of the order of 15. This strong correlation clearly indicates that the viscous sublayer has an important role in the operation of such devices, and was the motivation for a number of theoretical analyses [1]–[6] of the Stokes flow of a fluid past a ribletted (or grooved, depending on the way one sees it) surface.

In undertaking such analyses one must depart from the commonplace view of the viscous sublayer as a region where random fluctuations of the velocity field are totally absent. Indeed it is more appropriate to see the viscous sublayer as a region where the mean and fluctuating components of the velocity field superpose without interacting.

The viscous sublayer may be suitably defined as the region where the nonlinear convective terms of the Navier-Stokes equations are negligible and the viscous stresses alone determine the flow. The equations that govern the viscous sublayer are therefore linear and dissipative, so that no local instabilities arise. Nevertheless the flow may in general be driven unsteadily by the turbulent boundary layer above. Linearity, however, implies that the mean stress equals the stress calculated from the mean velocity field, whether oscillating components are present or not; this is why oscillations are usually unimportant.

That oscillations are indeed present may be understood by the following qualitative reasoning. Roughly speaking, oscillations governed by the Stokes equations only travel as far as a wavelength from their source. Although nonlinear sources are negligible inside the viscous sublayer, they are active just above, and in a large range of wavelengths: oscillations of a wavelength small compared to the viscous sublayer's height do not penetrate into it significantly, but oscillations of a wavelength large compared to the sublayer's height persist all the way down to the wall (mathematically, they behave as *quasi-steady* solutions of the Stokes equations [4]), with an amplitude that decreases linearly, just as the mean speed, and retains a constant ratio to it. Therefore, while wall features, such as riblets or wire arrays, acting inside the viscous sublayer cannot have a significant effect on turbulent structures, such as streamwise vortices, of a size small compared to the viscous sublayer itself, they can indeed affect structures larger than the viscous sublayer in size. Since it has been shown [3, 4] that riblets impede a quasi-steady crossflow more than a parallel flow, we may assume that wall structures embedded in the viscous sublayer provide some damping of long-wavelength streamwise vortices.

Unfortunately the nonlinear stage of the phenomenon, namely, the interaction of riblets with long-wavelength turbulent eddies producing a drag reduction, has until today only been explained qualitatively: all that we know quantitatively is that riblets appear to a transverse and to a longitudinal quasi-steady flow with different protrusion heights. Once that is established, however, it is just reasonable to expect that they produce some damping of streamwise vortices and thus a reduction of turbulent mixing and momentum transfer.

As long as flow about the riblets remains governed by the Stokes equations, the bigger the riblets are, the larger is the damping; beyond some size, however, the Stokes equations lose validity, nonlinearities come into play, and drag starts to grow again, thus giving rise to the experimentally observed minimum near $s^+ = 15$.

A similar reasoning may be applied to wire arrays, and therefore it is interesting to perform an analysis of Stokes flow past wire arrays. We shall do so in the continuation of this paper, with the aim of calculating the relevant protrusion heights and comparing them with those of riblets. Our expectation of a similarity between wire arrays and riblets has actually been confirmed by recent experiments performed after the completion of this work [9], which reported an optimum performance of wire arrays for a repetition period of $s^+ \simeq 15$ viscous units, just as for riblets.

In order to compare wire arrays with riblets on the grounds of their Stokes-flow characteristics, we need a single global parameter that may reasonably represent the overall performance of different riblet and wire-array profiles for drag reduction. Based on the above qualitative analysis, one such parameter may be constructed as the ratio of the protrusion-height difference to the longitudinal protrusion height, which we shall call "relative protrusion-height difference".

In fact, while the drag-reducing effects of riblets and related devices are proportional to the protrusion-height difference as long as the riblets are small enough, the eventual limitation to how big these devices may be made comes from nonlinear convective effects that set in once the size of the perturbed region increases beyond certain limits. A measure of nonlinear effects is given by the longitudinal protrusion height itself, because this quantity is proportional to the maximum velocity that would be found at the surface if the linear profile existing at infinity were continued down unperturbed. The corrugation shape that achieves the best balance between linear eddy damping and unwanted nonlinear effects may be expected to be the one that yields the maximum protrusion-height difference for a given longitudinal protrusion height; in other words, the dimensionless ratio of protrusion-height difference to longitudinal protrusion height is our performance parameter.

3. Longitudinal Stokes flow past an isolated wire array

In the limit of small wire diameter an analytical solution of the Stokes equation may be constructed using the technique of multipole expansion. In addition, this problem admits of a definite limit when the wall is pushed away indefinitely far from the wire screen. We shall start with this, simpler, limiting case.

As observed in Ref. [4], for a translationally invariant structure such as riblets or wire arrays the Stokes equations split into two separate problems: a Laplace equation for the longitudinal flow and a biharmonic equation for the transverse flow. In order to determine the longitudinal flow we have to solve the Laplace equation $\Delta_2 w = 0$ with the conditions that the longitudinal velocity w, periodic of period 2π, should vanish at the surface of each wire, seen in a transverse plane as a circumference of radius a, and should attain a unit gradient for y going to plus infinity.

In the limit of small a, the solution of this problem is simply given by the Green function of Ref. [4], Eq. (27):

$$G(x, y) = [\log(2\cosh y - 2\cos x) + y]/2 \tag{1}$$

In fact, $G(x, y)$ is periodic, satisfies the required condition at infinity, and behaves in proximity of the origin as $\log r$, where $r^2 = x^2 + y^2$ (as may easily be found out from the Taylor-series expansions of cos and cosh). Since the contour lines of G become more and more approximately circular as $r \to 0$, it is sufficient to subtract $\log a$ from G to satisfy the no-slip condition at a small circumference of radius a. We thus obtain

$$w(x, y) = [\log(2\cosh y - 2\cos x) + y]/2 - \log a \tag{2}$$

as the required solution. For $y \to \infty$, $w \simeq y - \log a$, and therefore the protrusion height (referred to a corrugation period of 2π) is simply $-\log a$, if calculated from the center of the wire, or

$$h_{\parallel} = -\log a + a \tag{3}$$

if calculated, consistently with the definition adopted for riblet geometries, from the uppermost point of the wire surface.

There is a remarkable difference with the case of a grooved surface: the protrusion height increases without bound for $a \to 0$, rather than attaining a finite limit. It should, however, be noted that the fluid velocity existing in the neighbourhood of the wires increases as well (for instance at the centre point between two consecutive wires, $x = \pi$

and $y = 0$, w equals $\log 2 - \log a$), so that the assumption of negligible nonlinear terms in the flow equations becomes less and less applicable as a decreases.

It is also noteworthy that w attains a nonzero finite limit for y going to minus infinity. All other solutions of the Laplace equation are infinite on both sides of the wire array and none is zero at infinity; the finite solution actually represents the limit of flow about a wire array adjacent to a plane wall when the wall is moved away to infinity.

The approximation of the solution (2) may be improved, for a given a, by a multipole-expansion technique. The function $\frac{1}{2}\log(2\cosh y - 2\cos x)$ is in complex notation the real part of

$$F(z) = \log[2\sin(z/2)] \tag{4}$$

where $z = x + iy$, and behaves for $z \to 0$ as $\log z$. It represents a periodic array of monopoles. The n-th derivative of this function behaves for $z \to 0$ as

$$F^{(n)}(z) \simeq (-1)^{n-1}(n-1)!z^{-n} \tag{5}$$

and represents a periodic array of $2n$-poles (i.e., field configurations that in a neighbourhood of the origin depend on the angular coordinate θ as $e^{in\theta}$). Boundary conditions imposed on a circular boundary are efficiently handled term-by-term if the solution is represented as a linear superposition of multipoles (i.e., a Fourier expansion with respect to θ); higher-order multipole coefficients turn out to be proportional to a^{2n}, thus effectively representing an expansion of the solution, and of related quantities such as the protrusion height, in powers of a.

We performed the above expansion up to order 6 in a; the resulting expression of $h_{\parallel}$ is

$$h_{\parallel} = -\log a + a - \frac{a^2}{4} + \frac{5}{288}a^4 - \frac{a^6}{576} + O(a^8) \tag{6}$$

4. Cross flow past an isolated wire array

In order to determine the cross-flow pattern one must solve the biharmonic equation for the streamfunction ψ. A general solution of this equation may be written in the form $\psi = f + yg$, where f and g are harmonic functions (solutions of the Laplace equation), which near a circular boundary may efficiently be expressed as periodic multipole expansions. The coefficients of each of these expansions, which in a neighbourhood of the origin appear as Fourier expansions with respect to the angular coordinate θ, may then be determined by imposing that the Fourier coefficients of both ψ and ψ_r vanish up to a certain order.

The first six terms of such a double multipolar expansion, which may be found to contain all the contributions to ψ of order up to a^2, are given by

$$\psi = y[\log(2\cosh y - 2\cos x) + y]/4 + Ay + B + C\log(2\cosh y - 2\cos x)/2 + D\frac{\sinh y}{2\cosh y - 2\cos x} + Ey\frac{\sinh y}{2\cosh y - 2\cos x} + \cdots. \tag{7}$$

On equating to zero the zeroth, first and second-order Fourier coefficients of ψ and ψ_r on a circumference with radius a and truncating all results to order a^2, we obtain a system from which the coefficients A, B, C, D, E will be determined as

$$A = -\frac{1}{2}\log a - \frac{1}{4} - \frac{a^2}{48} + O(a^4), \quad B = \frac{a^2}{4}\log a + \frac{a^2}{8} + O(a^4),$$
$$C = -\frac{a^2}{4} + O(a^4), \quad D = \frac{a^2}{4} + O(a^4), \quad E = -\frac{a^2}{2} + O(a^4). \tag{8}$$

It is interesting to observe that the coefficient A is the only one that contains terms of order smaller than a^2, so that if accuracy $O(a^4)$ is not required the calculation may be limited to this single coefficient and to the *first* (not zeroth) order Fourier component of the boundary conditions. (D must also be considered during the solution of the linear system of equations for the coefficients, but turns out to be of order a^2, and therefore negligible in the complete solution.)

From the behaviour at infinity to order a^2 of the complete solution we may easily determine the transverse protrusion height to the same order. Indeed, we continued the multipole expansion up to order a^6, thus obtaining:

$$h_\perp = -\frac{\log a}{2} - \frac{1}{4} + a - \frac{5}{12}a^2 + \frac{43}{576}a^4 - \frac{79}{3840}a^6 + O(a^8). \tag{9}$$

5. Longitudinal flow past a wire array parallel to a plane wall

In the presence of a plane wall located a distance h_w below the center of the wires, as in Fig. 1, the multipole terms that constitute the solution given in Section 3 must be modified to include the presence of images, in such a manner that the no-slip condition is satisfied at the wall. The solution to leading order has the form

$$w = y + h_w + A\left\{\frac{\log(2\cosh y - 2\cos x)}{2} - \frac{\log[2\cosh(y + 2h_w) - 2\cos x]}{2}\right\} + O(a^2). \tag{10}$$

Determining the coefficient A so that w vanishes, to zeroth order in its multipole expansion, on a small circumference with radius a, we obtain

$$A = -h_w/\{\log a - \log[2\cosh(2h_w) - 2]/2\}, \tag{11}$$

whence the protrusion height:

$$h_\parallel = (1 - A)h_w + a + O(a^2). \tag{12}$$

It may be verified that when $h_w \gg -\log a$ the solution (10) and the protrusion height (12) tend to those given in Section 3.

We have repeated the above calculations to order a^2, with the aid of a computerized symbolic processor. The resulting expression, too long to be reported here, has been used for the diagrams illustrated below.

6. Cross-flow past a wire array parallel to a plane wall

In the case of cross-flow (biharmonic equation) the procedure necessary to introduce the presence of a plane wall is formally similar to that adopted in the previous section, but the details are more complicated. According to the decomposition of the biharmonic solution in terms of two harmonic functions as $\psi = f + yg$, we need the "image", that is an additional term adjusting things so that the two components of velocity vanish at the wall, of both harmonic multipoles and harmonic multipoles multiplied by y.

Locating the wire at the origin and the plane wall at $y = -h_w$, we prefer to decompose the solution of the biharmonic equation as $\psi = f + (y + h_w)g$; the two boundary conditions at the wall then read

$$f(x, -h_w) = 0; \quad f_y(x, -h_w) + g(x, -h_w) = 0. \tag{13}$$

We must now distinguish the two cases of a monopole (or multipole) term appearing in f or in g. When a multipole component is assigned in g, f may be set identically equal to zero, and it is sufficient to satisfy the condition $g(x, -h_w) = 0$ by an image analogous to that used in the case dealt with in Section 5 to comply with Eq. (13). When, on the other hand, the multipole component is assigned in f, we must calculate both an f-image such that $f(x, -h_w) = 0$ and a g-image such that the second of Eq. (13) is also satisfied.

According to the above criteria, and to the considerations made in Section 4 as to the relevance of different multipole terms, we may write the solution, to leading order in the wire radius a, as

$$\psi = (y + h_w)\left(\frac{y + h_w}{2} + A\left\{\frac{\log(2\cosh y - 2\cos x)}{2} - \frac{\log[2\cosh(y + 2h_w) - 2\cos x]}{2}\right\}\right)$$
$$+ B\left\{\log(2\cosh y - 2\cos x)/2 - \log[2\cosh(y + 2h_w) - 2\cos x]/2 + (y + h_w)\frac{\sinh(y + 2h_w)}{\cosh(y + 2h_w) - \cos x}\right\} + O(a^2), \tag{14}$$

where the constants A and B (together with a dipole coefficient D which, being of order a^2, cannot legitimately be included in the solution (14) without also considering the contributions of order a^2 that come from other terms) must be determined by equating to zero the zeroth and first Fourier components of ψ_r and the first Fourier component of ψ on a small circumference with radius a. (The zeroth Fourier component of ψ is not zero; it equals the flow rate through the gap between wire and wall and is a result, rather than a datum, of the problem.)

In particular, the local approximation of Eq. (14) near the origin is given by

$$\psi = h_w^2/2 + h_w r\sin\theta$$
$$+ A\left\{h_w\left[\log r - \frac{\log(2\cosh 2h_w - 2)}{2}\right] - h_w\frac{\sinh(2h_w)}{2\cosh(2h_w) - 2}r\sin\theta + \left[\log r - \frac{\log(2\cosh 2h_w - 2)}{2}\right]r\sin\theta\right\}$$
$$+ B\left\{\log r - \frac{\log[2\cosh(2h_w) - 2]}{2} - \frac{\sinh(2h_w)}{2\cosh(2h_w) - 2}r\sin\theta + h_w\frac{\sinh(2h_w)}{\cosh(2h_w) - 1} - h_w\frac{1}{\cosh 2h_w - 1}r\sin\theta + \frac{\sinh(2h_w)}{\cosh(2h_w) - 1}r\sin\theta\right\} + Dr^{-1}\sin\theta + O(r^2), \tag{15}$$

and its derivative with respect to r by

$$\psi_r = h_w\sin\theta + A\left\{h_w r^{-1} - h_w\frac{\sinh(2h_w)}{2\cosh(2h_w) - 2}\sin\theta + [\log r + 1 - \log(2\cosh 2h_w - 2)/2]\sin\theta\right\}$$

$$+ B\left\{r^{-1} - \frac{\sinh(2h_w)}{2\cosh(2h_w) - 2}\sin\theta - h_w\frac{1}{\cosh 2h_w - 1}\sin\theta\right.$$

$$\left. + \frac{\sinh(2h_w)}{\cosh(2h_w) - 1}\sin\theta\right\} - Dr^{-2}\sin\theta + O(r). \tag{16}$$

On letting $r = a$ and equating to zero the coefficient of $\sin\theta$ in ψ and the whole of ψ_r we obtain the coefficients A, B, D and hence the protrusion height $h_\perp$.

Actually, just as in the case of longitudinal flow, the calculations have been performed to order a^2 with a symbolic processor, and the resulting expressions are too long to be reported here. The results are described below.

7. Analysis of the results

In Reference [4] the difference of the longitudinal and transverse protrusion heights was assumed as a parameter measuring the capability of a grooved surface to damp cross-flow vortices, on the basis that this difference is the only combination of the two protrusion heights that is independent of the choice of the origin of the coordinate system. The difference Δh was there made dimensionless by normalizing it to the period of the corrugations. It turned out that the absolute maximum value of the normalized protrusion-height difference, namely 0.132, was given by an array of infinitely thin and high blades.

In connection with wire arrays, a similar normalization of the protrusion-height difference leads to a startling result: $\overline{\Delta h}$ can be made arbitrarily large by reducing wire diameter. In fact, combining Eqs (3) and (10) gives

$$\overline{\Delta h} = -\frac{1}{4\pi}\log a + \frac{1}{8\pi} + O(a^2), \tag{17}$$

which is infinite for $a \to 0$.

However, it is easy to imagine that the conclusion that the performance of a wire-array device can be improved indefinitely by making the wires thinner and thinner is too good to be true. When wire diameter is decreased (with the period kept constant), the velocity of flow in proximity of the wires also increases indefinitely, so that the flow must eventually abandon the linear viscous regime. On the other hand, there is no apparent reason to keep the period fixed, and by changing the period one may increase at will the protrusion height of riblets as well.

As was mentioned in Section 2, the eventual limitation to how big riblets may be made, or how thin wires may be made, comes from nonlinear convective effects that set in once the region perturbed by their presence grows beyond certain limits. The parameter that qualitatively appears to represent a balance between linear eddy damping and harmful nonlinearities is the relative protrusion-height difference.

In Fig. 2 the relative protrusion-height difference is plotted versus riblet height for the three riblet profiles considered in reference [4]. As may be seen, the three curves are monotonically increasing, so that the main conclusion of that paper is unchanged: the best performance, that is the absolute maximum value of the relative protrusion-height difference, still corresponds to infinitely thin and high blades. This maximum is 0.598.

The relative protrusion-height difference of wire arrays, as obtained by a multipole expansion up to order a^2, is plotted versus wire diameter $\bar{d}$ in Fig. 3 for several values of wire-to-wall distance $\bar{h}_w$ (both $\bar{d}$ and $\bar{h}_w$ are expressed as a fraction of the period, so that $\bar{d} = a/\pi$ and $\bar{h}_w = h_w/2\pi$). The curve corresponding to $h_w = \infty$ is repeated in Fig. 4 for different orders of approximation in order to test the adequacy of the second-order multipole expansion. The cross-flow streamline pattern obtained in the particular example of $\bar{h}_w = 1$ and $\bar{d} = 0.1$ is also shown, in Fig. 5.

It may be noticed that $\Delta h/h_{\parallel}$, contrary to $\Delta h/2\pi$, stays finite when wire diameter tends to zero; in particular, when $h_w = \infty$ (the most favourable case) it goes to the limit 1/2, less than the value corresponding to infinitely deep grooves. $\Delta h/h_{\parallel}$ has a broad maximum of 0.589 for $h_w = \infty$ and a nonzero diameter of roughly 6–7% of the period. This value is still, even if marginally, lower than that corresponding to blade-style riblets. It may probably be improved slightly by making the wires pointed upwards rather than round-shaped.

The conclusion is that under optimal conditions the drag-reducing effects of wire arrays are likely to be similar to, but not better than, those of riblets. This conclusion

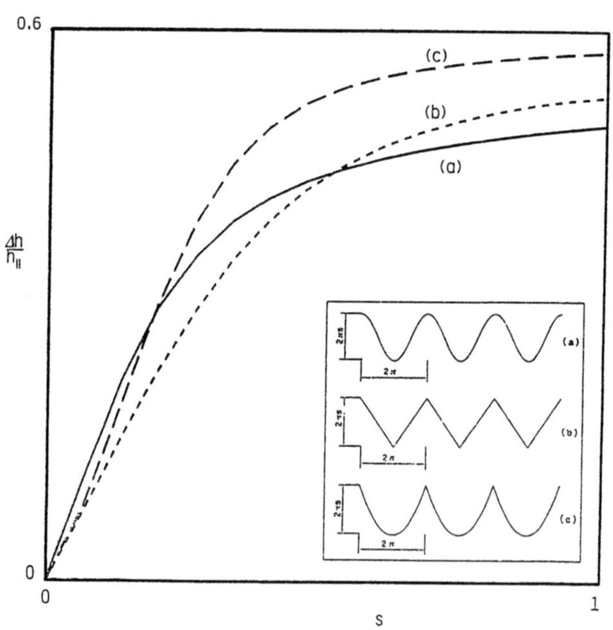

Fig. 2. Plot of the relative protrusion-height difference versus riblet height for several groove profiles.

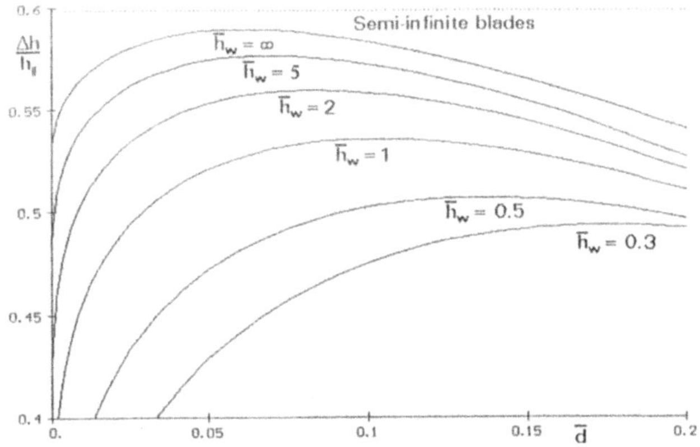

Fig. 3. Relative protrusion-height difference of wire arrays of varying diameter and height from wall.

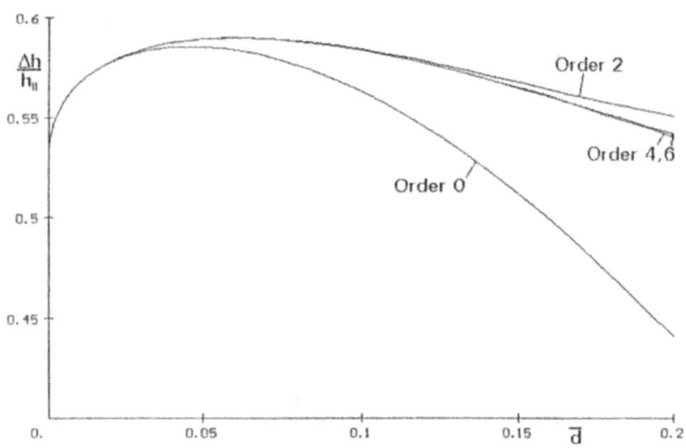

Fig. 4. Relative protrusion height difference of wire arrays isolated in space calculated at several orders of approximation in the multipole expansion.

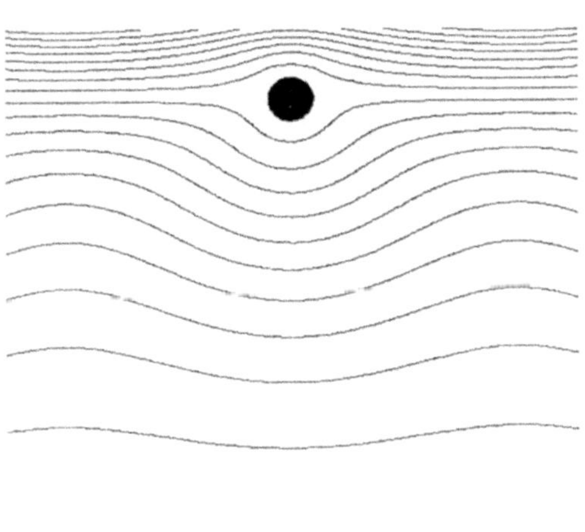

Fig. 5. Cross-flow streamline pattern in a wire array with $\bar{h}_w = 1$ and $\bar{d} = 0.1$.

appears to be confirmed by recent experiments [9] performed after the completion of this work.

Acknowledgement

This work was funded by the Italian Ministry of University and Research.

References

1. Bechert, D. W., Bartenwerfer, M. and Reif, W.-E., On the drag reduction of the shark skin. *AIAA* paper 85-0546 (1985).
2. Bechert, D. W. and Bartenwerfer, M., The viscous flow on surfaces with longitudinal ribs. *J. Fluid Mech.* 206 (1989) 105–129.
3. Bechert, D. W., Bartenwerfer, M. and Hoppe, G., Turbulent drag reduction by nonplanar surfaces – a survey on the research at TU/DLR Berlin. In: Gyr, A. (ed.), *Proc. IUTAM Symp. on Structure of Turbulence and Drag Reduction.* [Zürich, 25–28 July 1989.] Berlin: Springer (1990) pp. 525–543.
4. Luchini, P., Manzo, F. and Pozzi, A., Resistance of a grooved surface to parallel flow and cross-flow. *J. Fluid Mech.* 228 (1991) 87–109.
5. Luchini, P., Manzo, F. and Pozzi, A., Resistance of a grooved surface to parallel and cross flow calculated by B.E.M. In: Marino, L. and Piva, R. (eds), *Proc. IABEM (International Association for Boundary Element Methods) Symp.* [Rome, 15–18 Oct. 1990.] Berlin: Springer (1991) pp. 328–338.
6. Luchini, P., Manzo, F. and Pozzi, A., Viscous eddies in flow over a grooved surface computed by a Gaussian-integration Galerkin boundary-element method. *AIAA J.* 30 (1992) 2168–2170.
7. Bartenwerfer, M. and Bechert, D. W., Die viskose Strömung über behaarten Oberflächen. *Z. Flugwiss. Weltraumforsch.* 15 (1991) 19–26.
8. Kramer, M. O., Einrichtung zur Verminderung des Reibungswiderstandes. Reichpatentamt, Patentschrift Nr. 669 897, Klasse 62b, Gruppe 408 (1939). Patented from 17 March 1937 on.
9. Bechert, D. W., Hoppe, G., Hage, W. and Bruse, M., Widerstandsvermindernde Oberfläschen der Natur im Experiment. 1. Bionik-Kongress, Wiesbaden, 1–13 June 1992.

Applied Scientific Research **50**: 267–282, 1993.
K. Krishna Prasad (ed.), Further Developments in Turbulence Management.

Riblet flow calculation with a low Reynolds number $k - \varepsilon$ model

L. DJENIDI AND R.A. ANTONIA

Department of Mechanical Engineering, University of Newcastle, N.S.W., 2308, Australia

Abstract. A low Reynolds number $k - \varepsilon$ model has been used to calculate the turbulent boundary layer over riblets. Calculated mean velocity, Reynolds shear stress and kinetic energy distributions are generally in good agreement with available experimental data. The comparison between these distributions and those in a corner flow points to a significant difference between the two flows and the unlikelihood of counter-rotating vortices within the riblet grooves. One shortcoming of the present $k - \varepsilon$ model is the relatively slow return to a two-dimensional turbulence state as the distance from the riblet surface increases.

1. Introduction

A significant amount of research, perhaps starting with the work of Walsh and Weinstein (1978) has established that riblets (longitudinal wall grooves) can reduce the drag in a turbulent boundary layer by about 7–8%. However, despite this, the manner in which riblets interact with the turbulence structure is not yet understood. The existence of low-speed fluid inside the riblet valleys is fairly well established, thus pointing to the importance of viscous effects in this flow region (e.g. Choi et al., 1991) and a possible reason for the drag reduction (Djenidi et al., 1992). Djenidi et al. (1992) made extensive experimental and numerical studies of a laminar boundary layer over V-grooves and found that, despite a large increase in the wetted surface (100% when $s = h$; s is the distance between adjacent peaks and h is the height of the groove), the skin friction is at worst equal to that on a smooth plate. They conclude that viscous effects should play an important role in the drag reduction mechanism. When the flow is turbulent, the state of knowledge is much less satisfactory than in the laminar case, partly because of the dearth of experimental information within the grooves and partly because the available data in the wall region above the grooves suffer from a lack of accuracy. Consequently, our understanding of the drag reduction mechanism is poor and our ability to develop a reliable numerical model is severely hampered.

An early numerical investigation of turbulent boundary layer over a riblet surface was done by Khan (1986), with a mixing length-eddy viscosity approach to model the Reynolds shear stress. He solved the x-momentum and vorticity equations. Although a reasonable result was obtained for the drag reduction, the calculation revealed a rather unrealistic secondary flow. Later, Djenidi et al. (1991) used also a mixing length-eddy viscosity model and solved the x-momentum equation using a curvilinear mesh. Although the calculated velocity profiles were in good agreement with experiment (Vukoslavcevic et al., 1987) the drag reduction was over-estimated. Benhalilou et al. (1991) also used a mixing length approach, solving the x-momentum equation in a cartesian mesh. Their numerical results, despite being in qualitative agreement with their own experiments, overestimated the drag reduction. It would therefore appear that the mixing length-eddy viscosity model is inadequate for calculating the turbulent boundary layer over riblets and a more refined closure is required. The general aim of the present investigation was to develop a transport model for calculating the skin friction and distributions of the mean velocity, Reynolds stress and kinetic energy in a turbulent boundary layer over riblets. As a first step towards this goal, a low Reynolds number $k - \varepsilon$ model (Chien, 1982) was used in this paper. The resulting calculations are compared with available measurements.

2. Mathematical formulation

The equations embodying the mathematical model can be written in any of several co-ordinate systems. In the present study, the problem is solved using an orthogonal curvilinear system. Details of the derivation of the conservation equations in general orthogonal curvilinear co-ordinates may be found in Nash and Patel (1978).

2.1. Computational Grid

The curvilinear mesh used to compute the flow was derived using a Schwarz-Christoffel transformation: the computational mesh, Fig. 1a, is transformed into the physical mesh, Fig. 1b (see Bechert and Bartenwerfer, 1989). The V-groove shape is considered here (the flow direction is perpendicular to the $y - z$ plane; only half a riblet is shown). The smooth wall is simulated when θ, the groove half-angle, is 90°. The upper limit is located at about 10 times the boundary layer thickness—note that

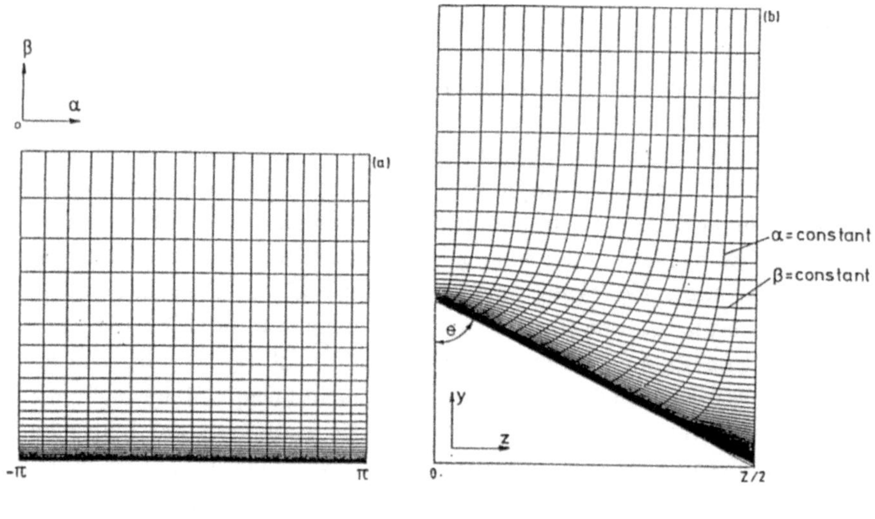

Fig. 1. (a) Computation and (b) physical mesh for half a V-groove.

only the region up to $y \simeq 3h$ is shown in Fig. 1. A series of numerical experiments established that the calculation was not particularly sensitive to the grid spacing, a 81×19 grid yielding the same $\bar{U}$, $\overline{uv}$ and k distributions as a 81×43 grid. It should be noted that despite the rather inevitably coarse spanwise spacing in the trough, numerical inaccuracies in this region have a negligible influence, particularly on the total skin friction since only less than about 1% of this friction is associated with this region for a 81×19 grid with $s = 0.7$ mm and $h = 0.35$ mm. Further, there are about 30 grid points in the y direction within the riblet ($h^+ \simeq 15$) above the trough, the first point being at $y^+ = 0.55$ for $R_\theta = 7700$.

The computational domain is defined by the riblet wall, over a 1 m length, an upper limit at 10δ and the lateral boundaries at $z = 0$ and $z = s/2$.

2.2. Mass and mean momentum conservation equations

Starting from the ensemble averaged Navier-Stokes equations, the stationary boundary layer assumptions were applied with $\partial P/\partial x = 0$. Thus the equations are

$$\frac{\partial h^2 U}{\partial x} + \frac{\partial h V}{\partial \beta} + \frac{\partial h W}{\partial \alpha} = 0, \tag{1}$$

$$\frac{\partial h^2 UU}{\partial x} + \frac{\partial hUV}{\partial \beta} + \frac{\partial hUW}{\partial \alpha} = \frac{\partial}{\partial \beta}\left(\nu \frac{\partial U}{\partial \beta} - h\overline{uv}\right) + \frac{\partial}{\partial \alpha}\left(\nu \frac{\partial U}{\partial \alpha} - h\overline{uw}\right), \tag{2}$$

where U, V, W, u, v and w are the components of the mean velocity and their fluctuations in the x (streamwise), β and α directions respectively, and the local metric h is a function of β, α, y and z (the normal and spanwise directions in a cartesian system) [see Djenidi et al., 1991]. ν is the kinematic viscosity of the fluid. The equations have five unknowns, U, V, W, $\overline{uv}$ and $\overline{uw}$. The closure of $\overline{uv}$ and $\overline{uw}$ will be considered in Section 2.3. Equations (1) and (2) are solved for U, and V. To obtain W, it is assumed that the mean streamwise vorticity is zero (everywhere in the flow), viz.

$$\Omega_x = \frac{1}{h^2}\left(\frac{\partial hW}{\partial \beta} - \frac{\partial hV}{\partial \alpha}\right) = 0. \tag{3}$$

The following evidence can be given in support of (3). It should first be noted that the riblets are largely immersed within the inner region ($h^+ \lesssim 20$) and the Reynolds stresses within the grooves in this region are unlikely to be of sufficient strength to generate secondary flows, e.g. counter-rotating vortices. Also, the relatively few detailed experimental data that are available (Hooshmand et al., 1983; Vukoslavcevic et al., 1987; Benhalilou et al., 1991; Benhalilou, 1992) do not seem to support such motions. In spite of the strong inhomogeneity in the wall region, the mean flow recovers its two-dimensionality ($W \equiv 0, \partial/\partial z \equiv 0$) for $y^+ \gtrsim 10$ above the riblets. On the basis of his measurements, Benhalilou (1992) concluded that there were no counter-rotating vortices within the grooves.

2.3. Turbulence modeling

An eddy viscosity, ν_t, is used to model the Reynolds stress, viz.

$$\overline{uv} = -\nu_t \frac{\partial U}{h\partial\beta}, \qquad \overline{uw} = -\nu_t \frac{\partial U}{h\partial\alpha}$$

with

$$\nu_t = C_\mu \frac{k^2}{\varepsilon}. \tag{4}$$

The turbulent kinetic energy k and the "isotropic" dissipation ε, are evaluated by a low Reynolds number $k-\varepsilon$ model (Chien, 1982). In the present curvilinear co-ordinates, the equations are

$$\begin{aligned}&\frac{\partial}{\partial x}(h^2Uk) + \frac{\partial}{\partial\beta}(hVk) + \frac{\partial}{\partial\alpha}(hWk)\\ &\qquad = h^2P_k - h^2(\tilde{\varepsilon} + D) + \frac{\partial}{\partial\beta}\left[\left(\nu + \frac{\nu_t}{\sigma_k}\right)\frac{\partial k}{\partial\beta}\right] + \frac{\partial}{\partial\alpha}\left[\left(\nu + \frac{\nu_t}{\sigma_k}\right)\frac{\partial k}{\partial\alpha}\right]\end{aligned} \tag{5}$$

$$\begin{aligned}&\frac{\partial}{\partial x}(h^2U\tilde{\varepsilon}) + \frac{\partial}{\partial\beta}(hV\tilde{\varepsilon}) + \frac{\partial}{\partial\alpha}(hW\tilde{\varepsilon})\\ &\qquad = C_{\varepsilon_1}f_1\frac{\tilde{\varepsilon}}{k}h^2P_k - C_{\varepsilon_2}h^2f_2\frac{\tilde{\varepsilon}^2}{k} + \frac{\partial}{\partial\beta}\left[\left(\nu + \frac{\nu_t}{\sigma_\varepsilon}\right)\frac{\partial\tilde{\varepsilon}}{\partial\beta}\right]\\ &\qquad\quad + \frac{\partial}{\partial\alpha}\left[\left(\nu + \frac{\nu_t}{\sigma_\varepsilon}\right)\frac{\partial\tilde{\varepsilon}}{\partial\alpha}\right] + E\end{aligned} \tag{6}$$

with $P_k = -h\overline{uv}\,\partial U/\partial\beta - h\overline{uw}\,\partial U/\partial\alpha$, $\tilde{\varepsilon} = \varepsilon - D$. The definitions of D and E and the magnitudes of the constants C_μ, C_{ε_1}, C_{ε_2}, and σ_ε are given in Table 1a. The damping functions f_1, f_2 and f_μ (discussed below) are shown in Table 1b.

A low Reynolds number $k-\varepsilon$ model is an extension of the classical $k-\varepsilon$ model developed for high Reynolds numbers, to account for the near-wall effect. In this present model, ε is replaced by $\tilde{\varepsilon}$ which, by contrast to ε, goes to zero at wall. As regards comments on E, D and $f_i (i = 1, 2)$, the reader will find a complete description in Chien's (1982) paper. To reduce the eddy viscosity due to the near-wall effects,

Table 1a

D	E	C_μ	C_{ε_1}	C_{ε_2}	σ_k	σ_ε
$\frac{2\nu k}{y^2}$	$-\frac{2\nu\varepsilon}{y^2}\exp(-0.5y^+)$	0.09	1.35	1.8	1.0	1.3

Table 1b. $R_t = k^2/\nu\tilde{\varepsilon}$; $y^+ = U_\tau y/\nu$

f_1	f_2	f_μ
1.0	$1 - 0.22\exp\left[-\left(\frac{R_t}{6}\right)^2\right]$	$1 - \exp(-0.0115y^+)$

Chien (1982) modified Eq. (4) by including a damping function, f_μ

$$\nu_t = C_\mu f_\mu \frac{k^2}{\varepsilon}.$$

Chien's model was chosen partly because of the ease with which it can be implemented, partly because of its ability in reproducing the most basic features of a flat plate boundary layer and finally because of its accuracy in calculating the skin friction (Patel et al., 1985). This latter feature is important in the context of estimating the drag reduction due to riblets. Rodi and Mansour (1990) evaluated a number of $k - \varepsilon$ models against channel flow and boundary layer DNS data. They found that while the Chien model was not as accurate close to the wall as the model that they developed, it was more accurate by comparison to the Launder-Sharma (1974) or Lam-Bremhorst (1981) models.

2.4. *Initial and boundary conditions*

To start the calculations, initial mean velocity, turbulent kinetic energy and dissipation distributions are required. They were obtained from curve fits to the available data (experimental or numerical) or from theoretical formulations. Preliminary calculations showed a negligible effect of the precise form of the initial distributions on the results. The boundary conditions were:

at the wall	$U = V = W = k = \tilde{\varepsilon} = 0$
at the upper limit of the computational domain	$\partial/\partial y = 0$
on the lateral boundaries ($z = 0$, $z = s/2$)	$\partial/\partial z = 0.$

2.5. Numerical method

A control finite volume procedure developed by Patankar (1980), is used for the discretisation of the equations (1–6), based on an orthogonal mesh and employing a staggered grid where the V and W nodes are shifted from the U, k and $\tilde{\varepsilon}$ nodes. Integration of each equation of transport across the appropriate control volumes yielded finite volume equations of the form, i.e.

$$(A_p - S_1)\phi_p = A_N\phi_N + A_s\phi_s + A_E\phi_E + A_W\phi_W + S_2 \tag{7}$$

where $A_p = A_N + A_s + A_E + A_W$ and S_1 and S_2 are coefficients of the linearised source terms. The A coefficients contain the combined effects of convection and a standard hybrid differencing scheme (Patankar, 1980). The solution procedure is an x-marching method, where the finite volume scheme was used at each step in the x direction (de St. Victor, 1986). The discretised system (7) was solved with the Modified Strongly Implicit algorithm (MSI) of Schneider and Zedan (1981).

3. Results

3.1. Flat plate

In order to test the equations written in a curvilinear co-ordinate system, calculations were performed on a flat plate (simulated by $\theta = 90°$, see Fig. 1) for $R_\theta = 1100$ and 7700. In Fig. 2, the mean velocity, $\bar{U}$, the Reynolds stress, $\overline{uv}$, and the turbulent kinetic energy, k, distributions are reported. There is close agreement between the present calculations at $R_\theta = 1100$, the measurements of Vukoslavcevic et al. (1987) at $R_\theta = 1000$ and Spalart's (1988) DNS data for $R_\theta = 1410$. There is also good agreement between the present calculations at $R_\theta = 7700$ and Klebanoff's (1955) measurements ($R_\theta = 7700$).

The above results indicate that no significant error has been introduced by transforming the equations of motion and closure model equations from a cartesian co-ordinate system to an ortho-curvilinear one, a transformation which necessitates the use of the scale factor h in all the equations. It is also satisfying to observe that the model, originally developed for high values of R_θ seems to give satisfactory results at a relatively small value of R_θ. This is rather attractive in the present context since, in order to obtain detailed measurements on riblet flows, experiments have to be carried out with relatively large grooves and therefore at low Reynolds numbers (Vukoslavcevic et al., 1987, $R_\theta \simeq 1000$; Benhalilou et al., 1991, $R_\theta \simeq 300$).

3.2. Riblet wall

Before presenting the riblet results, it is perhaps important to explain the choice of the present curvilinear transformation. Previous calculations in laminar riblet flows

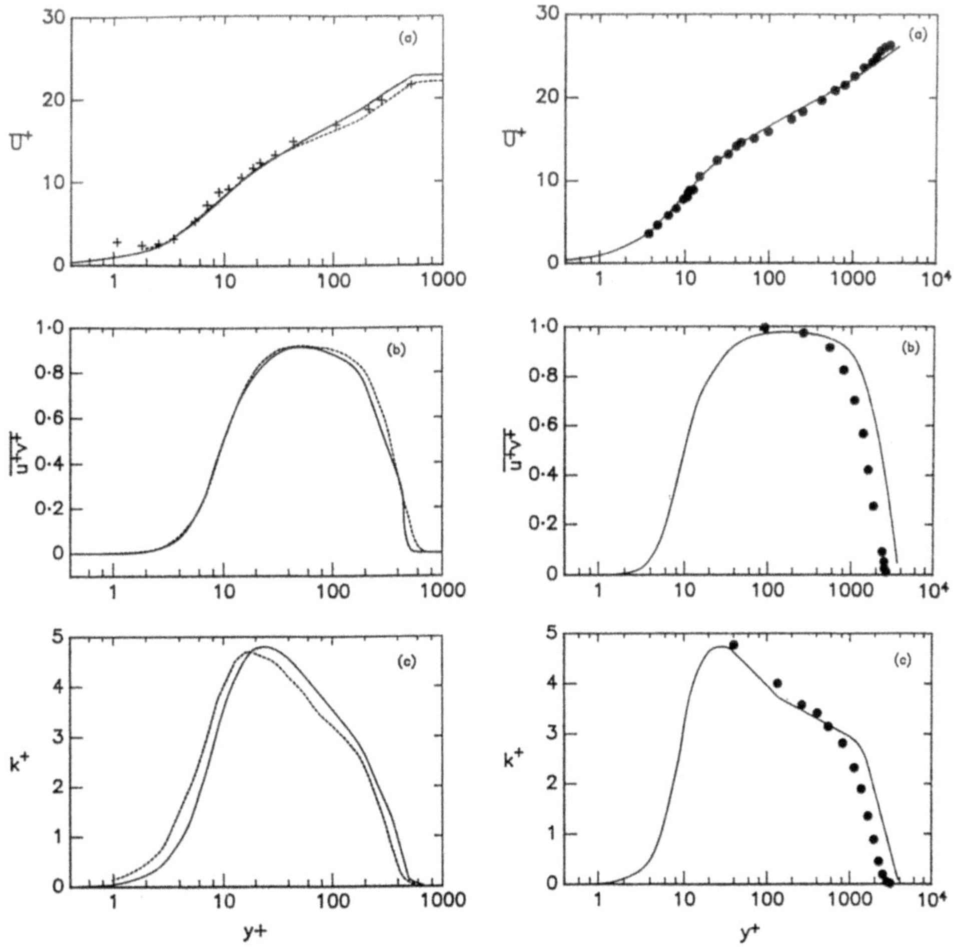

Fig. 2. Distributions of mean velocity, Reynolds shear stress and kinetic energy in a turbulent boundary layer over a smooth flat plate. (1) $R_\theta = 1000$; (2) $R_\theta = 7700$. ——, present calculation; +, Vukoslavcevic et al. (1987); – – –, Spalart (1988); ●, Klebanoff (1955).

(Djenidi et al., 1991) showed that an exactly 2-D flow ($W = 0$; $\partial/\partial z = 0$) could be achieved with this particular curvilinear system (Fig. 1). Contours of constant β and constant U (Fig. 8 of Djenidi et al., 1991) were identical. Bechert and Bartenwerfer (1989) also obtained the same result by theoretically solving

$$\Delta U = 0(\Delta U = \partial^2 U/\partial y^2 + \partial^2 U/\partial z^2)$$

for turbulent riblet flow: they assumed that since the riblets are immersed in the inner region of a turbulent boundary layer ($h^+ \leqslant 15 - 20$) the convective terms and the Reynolds shear stress are negligible in the equations of the motion. These results would show that the present co-ordinate system would be well adapted to the assumption $\Omega_x = 0$ [Eq. (3)]. Furthermore, in this curvilinear system, it is convenient to define the distance to the wall along β despite its ambiguous physical meaning (see Djenidi et al., 1991). This avoids using a distance to the wall as defined by Buleev (1963) and as is often done when dealing with walls (e.g. corners) that have spanwise inhomogeneities.

3.2.1. Mean velocity

The calculated mean velocity profiles $R_\theta \simeq 1100$ are compared in Fig. 3 with the experimental data of Vukoslavcevic et al. (1987) at $R_\theta \simeq 1000$ and of Benhalilou et al. (1991) at $R_\theta \simeq 300$. Note that the origin of the velocity profile above a peak has been shifted by a distance h^+. Because of the nature of the conformal mesh, especially in the vicinity of the wall, it is best to compare experiments and calculations above peaks and valleys. The agreement is good when R_θ is about the same ($R_\theta \simeq 1000$). In the region $y^+ \geqslant 15$, the difference between the calculations and the $R_\theta = 300$ measure-

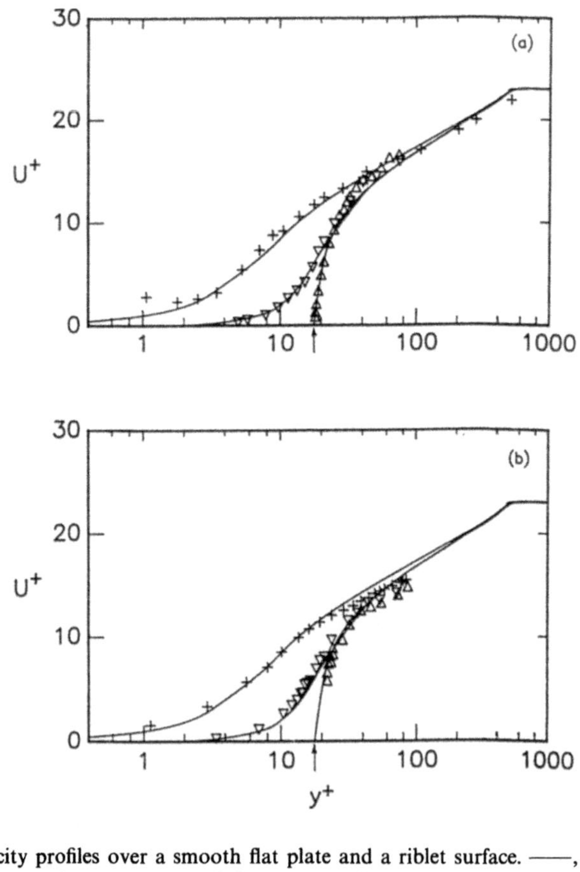

Fig. 3. Mean velocity profiles over a smooth flat plate and a riblet surface. ——, present calculation at $R_\theta = 1000$; +, ∇, Δ, experiments (+, smooth flat plate; Δ, riblet peak; ∇, riblet trough). The arrow indicates the location of the riblet peak; the riblet trough is at $y = 0$. (a) $R_\theta = 1000$ (Vukoslavcevic et al., 1987); (b) $R_\theta = 300$ (Benhalilou et al., 1991).

ments can probably be assigned to low Reynolds number effects (e.g. Spalart, 1988). The agreement for $y^+ \leqslant 15$ is consistent with the observation that $\bar{U}$ is relatively uninfluenced by the Reynolds number in this flow region (Spalart, 1988; Antonia et al., 1992; Wei and Willmarth, 1989; Djenidi and Antonia, 1992).

Beyond $y^+ \simeq 15$, the profiles above the riblet peak coincide with those over the riblet valley. This is true for both measurement and calculation.

3.2.2. *Reynolds shear stress and turbulent kinetic energy*

Figures 4 and 5 show the calculated profiles of $\overline{u^+v^+}$ and k^+ over a smooth flat plate above a riblet peak and a riblet trough for $R_\theta = 1000$ and 7700. Qualitatively $\overline{u^+v^+}$ is similar at both Reynolds numbers; the same observation applies to k^+. Further, both quantities exhibit identical behaviours within and just above ($y^+ \leqslant 15$) the groove.

Above the riblets ($y^+ \gtrsim 20$), $\overline{u^+v^+}$ is significantly different from k^+. As y^+ increases above the riblet surface, the present calculations of $\overline{u^+v^+}$ are unlikely to be correct. For example, by comparison to the unpublished data of Benhalilou (1992) [these are

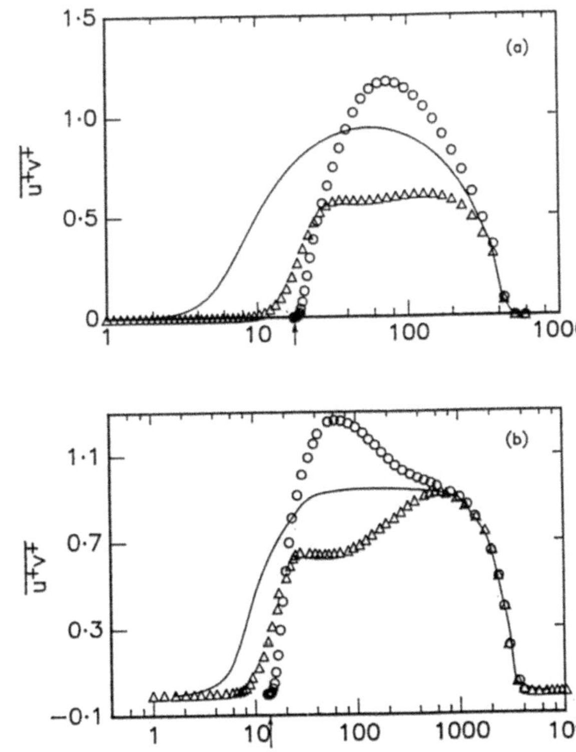

Fig. 4. Distributions of Reynolds shear stress. Calculations at (a) $R_\theta = 1000$ and (b) $R_\theta = 7700$. The arrow indicates the location of the riblet peak; the riblet trough is at $y = 0$. ——, smooth flat plate; ○, riblet peak; Δ, riblet trough.

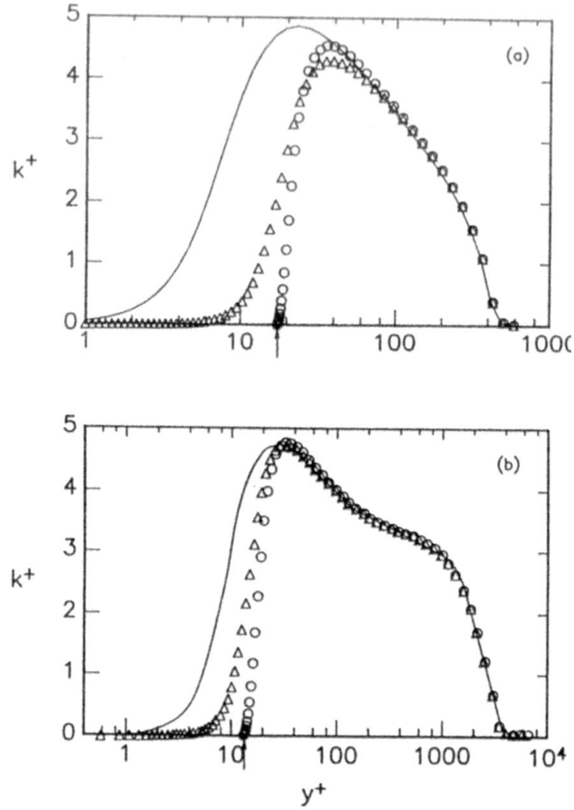

Fig. 5. Distributions of kinetic energy. Calculations at (a) $R_\theta = 1000$ and (b) $R_\theta = 7700$. The arrow indicates the location of the riblet peak; the riblet trough is at $y = 0$. ——, smooth flat plate; ○, riblet peak; Δ, riblet trough.

not shown here], it would appear that the calculation overestimates $\overline{u^+v^+}$ above a peak and underestimates it above a trough. This is probably caused by the manner in which the near-wall effect is taken into account by the closure model. Specifically, the present relation for f_μ does not appear to reproduce the apparent return to a two-dimensional state as y^+ increases. The use of U_τ in $y^+ = yU_\tau/\nu$ may not be appropriate here since it varies significantly along the groove contour (e.g. Fig. 6 where $U_{\tau o}$ is the friction velocity on the smooth plate.) Since $\overline{u^+v^+}$ depends strongly on $f_\mu(\overline{uv} = C_\mu f_\mu k^2/\varepsilon \partial U/\partial y)$, its spanwise variation may be incorrectly reproduced as the wall distance increases. This difficulty can be avoided by using a transport equation for $\overline{uv}$ where no damping function is required.

The mean velocity, Reynolds shear stress and turbulent kinetic energy distributions exhibit a common feature: they are all strongly damped within the groove, emphasising the importance of viscous effects on the near-wall turbulence. There is an apparent tendency towards "relaminarisation" within the grooves as shown in Fig. 7 ($\nu_t/\nu \leqslant 0.4$ everywhere within the riblets).

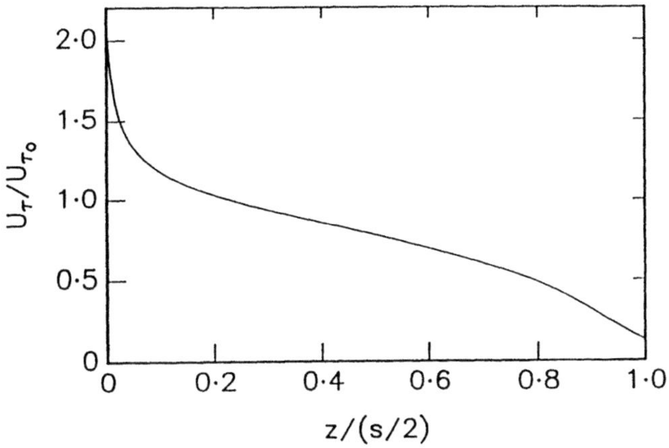

Fig. 6. Variation in friction velocity U_τ along the riblet contour. $s = 2h = 0.7$ mm; $R_\theta = 7700$.

3.2.3. Comparison between riblet and corner flows

It is of interest to compare the flow over riblets with that in a corner flow because of the geometrical similarity between the two cases, at least when V and L-shaped grooves are used. Constant velocity contours around one groove (Fig. 8a) can be compared with the corresponding contours for a corner (Fig. 8b, Bragg, 1969). As expected, the flow within the riblets behaves similarly to that in a corner, e.g. the spreading of the iso-contours as the corner is approached, suggesting a low skin

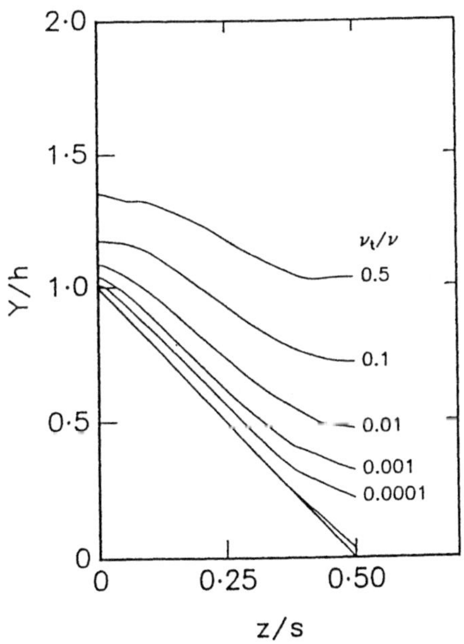

Fig. 7. Contours of constant eddy viscosity over riblets.

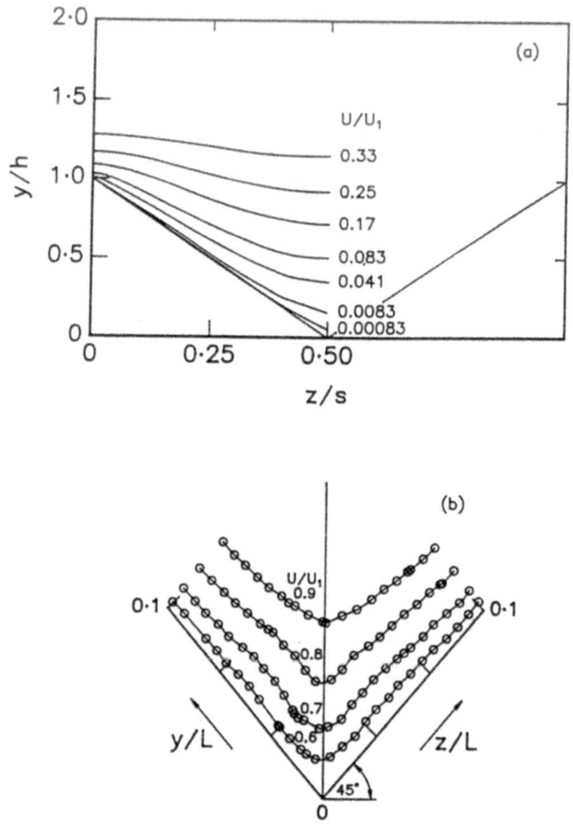

Fig. 8. Contours of constant mean velocity over riblets and in a corner flow. (a) riblet (calculation); (b) corner (experiment), L is the width of the plates (Bragg, 1969).

friction region. There is however a fundamental difference between the two flows in that the riblet height is small compared to the boundary layer thickness, δ, while the corner dimensions are comparable to δ. Accordingly, features of corner flows, such as the secondary motion would not be appropriate for riblet flows. Besides, a comparison between the Reynolds stress distributions along the corner bisector, Fig. 9, (Nakamura et al., 1982) and on the riblet trough (Figs 4 and 5) underlines the difference between the two flows: the double peaks observed on the $\overline{uv}$ and $\overline{u^2} + \overline{v^2} + \overline{w^2}$ $(=2k)$ in the corner flow due to the counter rotating motions are absent in the riblet flow.

3.2.4. *Skin friction and drag reduction over riblets*

The present calculations were carried out at $R_\theta \simeq 7700$. Fig. 10a shows the cumulative frictional drag along the riblet surface, $\int_0^l \tau \, dl$ (where $\tau = \mu \partial \bar{U}/\partial w|_{\text{wall}}$) normalised by the total friction over the riblet, $\int_0^L \tau \, dl$. Although the wetted surface area is greatly enhanced relative to that of a smooth surface, the major contribution to the total

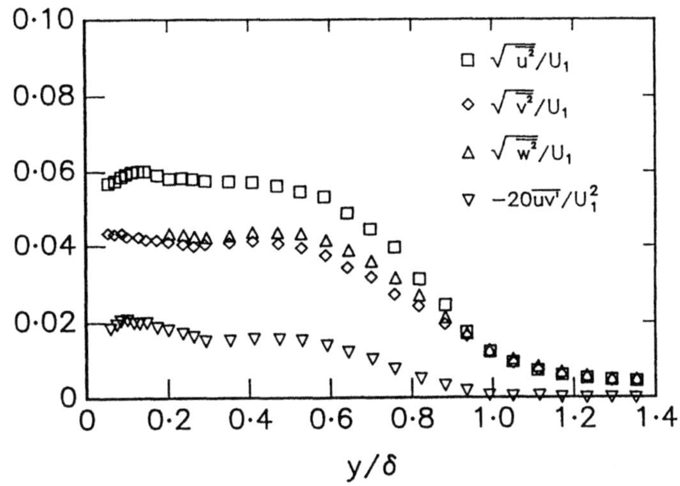

Fig. 9. Reynolds stress distributions along a corner bisector (Nakamura et al., 1982).

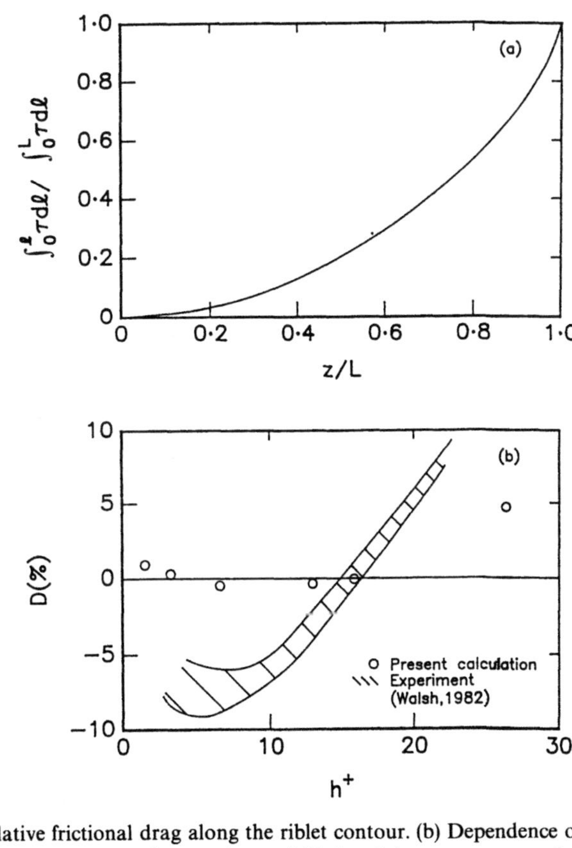

Fig. 10. (a) Cumulative frictional drag along the riblet contour. (b) Dependence on h^+ of the ratio of the total frictional drag over riblet surface to the total frictional drag over a smooth flat plate.

friction is associated with the region near the riblet peak. This suggests that the optimum riblet geometry should be one which minimises the frictional increase near the peak. The two most important parameters are therefore likely to be s/h and h/δ (or s/δ). However, since the riblet dimensions are comparable to the extent of the inner region, h/δ should be replaced by h^+.

The calculated frictional drag ratio, $D = 1 - F_R/F_0$, where F_R is the riblet frictional force and F_0 is the frictional force on a smooth flat plate, is shown in Fig. 10b for several values of h^+ (with $s^+ = 2h^+$). The calculated drag reduction is underestimated by comparison to the measurements of Walsh (1982). Note also that, when $h^+ \to 0$, D exceeds 0 whereas it should approach 0, as indicated by the experimental data. Launder and Li (1991) obtained a similar result for the calculated drag in a turbulent channel flow over riblets using another version of a low Reynolds number $k - \varepsilon$ model (Launder and Sharma, 1974). Low Reynolds number $k - \varepsilon$ models in their present two-dimensional forms clearly fail to simulate the spanwise inhomogeneity of the turbulent flow field over riblets.

4. Conclusions and final discussion

A low Reynolds number $k - \varepsilon$ model has been used to calculate a turbulent boundary layer over riblets. The calculated mean velocity is in agreement with available experiments. However, although the calculated turbulence quantities seem to be consistent with the measurements, the model does not reproduce the experimentally observed drag reduction.

The effect of riblets was limited mainly to the inner flow region, highlighting a fundamental difference between riblet and corner flows, where the characteristic length scale of the corner is usually comparable to the boundary layer thickness. This difference may support an argument against the existence of counter-rotating vortices inside the riblet grooves.

The present form of the $k - \varepsilon$ model is inadequate for capturing the disappearance of the spanwise variation in the turbulence quantities as the distance from the wall increases. In particular, the present damping function formulation reproduces the return towards two-dimensionality only poorly. While the introduction of the damping function is necessary, the choice of f_μ has a significant effect on the calculation of k and ε (Mansour et al., 1989), the modelling of k and ε may need to be reassessed (e.g. terms D and E in the present case). This may in turn require the various empirical constants $(C_\mu, C_{\varepsilon_1}, C_{\varepsilon_2}, \sigma_k, \sigma_\varepsilon)$ as well as the functions f_1 and f_2 to be modified. The damping effect on $\overline{uv}$ may be better simulated by using $\overline{v^2}/k$ instead of $f_\mu(\overline{uv} = C(\overline{v^2}/k)(k^2/\varepsilon)\partial U/\partial y)$, the rationale being that the rapid reduction in $\overline{uv}$ as the wall is approached is mainly due to the damping of $\overline{v^2}$ rather than the viscous effect (Launder and Tselipidakis, 1990). The attraction of this approach is that it avoids the use of U_τ, despite the necessity of introducing an equation for $\overline{v^2}$. Using this approach, good agreement with DNS smooth wall channel flow data was obtained by Durbin

(1990) and Kawamura (1991). The development of a Reynolds stress closure for riblet flow calculations should also be worthwhile despite the added complications to the equations. Its main advantage would be to avoid the use of empirical functions for simulating the near-wall effect.

Independently of whether the closure is of the $k - \varepsilon$ or Reynolds stress types, nonlocal parameters, such as the wall friction velocity, should not appear in the model. This would allow the spanwise variation in turbulence quantities to disappear quickly, thus permitting a faster relaxation towards two-dimensionality away from the riblet surface. An alternative way of avoiding the inhomogeneity of the non-local parameters, is to treat riblets as roughness elements and use calculation methods developed for rough surfaces (e.g. Granville, 1992). However attractive such a method may appear, it has the disadvantage of not addressing the physics behind the drag reduction.

A possible physical mechanism for drag reduction by riblets appears to be the prevention of the interaction between near-wall quasi-streamwise vortices and the riblet surface (Kim, 1992). In this context it is unlikely that $k - \varepsilon$ or Reynolds stress closures can capture the observed drag reduction since neither the characteristics of the vortices nor their interaction with the surface enter these models.

Acknowledgement

The support of the Australian Research Council is gratefully acknowledged.

References

Antonia, R.A., Teitel, M., Kim, J. and Browne, L.W.B., Low Reynolds number effects in a fully developed turbulent duct flow. *J. Fluid Mech.* 236 (1992) 579–605.

Bechert, D.W. and Bartenwerfer, M., The viscous flow on surfaces with longitudinal ribs. *J. Fluid Mech.* 206 (1989) 105–129.

Benhalilou, M., Etude Experimentale et Numerique d'une Couche Limite Turbulente sur Paroi Rainuree, Ph.D. Dissertation, I.M.S.T., France (1992).

Benhalilou, M., Anselmet, F., Liandriat, J. and Fulachier, L., Experimental and numerical investigation of a turbulent boundary layer over riblets. *Turbulent Shear Flows* 8 Munich (1991).

Bragg, G.M., The turbulent boundary layer in a corner. *J. Fluid Mech* 36 (1969) 485–503.

Buleev, N.I., Theoretical model of the mechanism of turbulent exchange in fluid flows. *AERE Trans.* 957, Harwell (1963) 1–39.

Chien, K.Y., Predictions of channel and boundary-layer flows with a low-Reynolds number turbulence model. *AIAA Jnl.* 20 (1982) 33–38.

Choi, H., Moin, P. and Kim, J., On the effect of riblets in fully developed laminar channel flows. *Phys. Fluids A* 3 (1991) 1892–1896.

de St. Victor, X., Resolution des Equations de Navier-Stokes Bi- ou Tridimensionelles par Méthodes de Marches—Application au Calcul de Melange d'Ecoulements Cisailles, Ph.D. Dissertation, I.N.P.T., France (1986).

Djenidi, L. and Antonia, R.A., LDA measurements in low Reynolds number turbulent boundary layer. *Expts. in Fluids* [to appear].

Djenidi, L., Liandriat, J., Anselmet, F. and Fulachier, L., Laminar boundary layer over riblets. *Phys. Fluids A* [submitted].

Djenidi, L., Squire, L.C. and Savill, A.M., High resolution conformal mesh computation for V, U and L-groove riblets in laminar and turbulent boundary layer. In: K. S. Choi (ed.), *Recent Development in Turbulence Management*. Dordrecht: Kluwer, Academic Publishers (1991) pp. 65–92.

Durbin, P.A., Near wall turbulence closure modeling without damping function. *CTR Manuscript* 112, Stanford University (1990).

Granville, P.S., Similarity-law analysis and turbulence modeling for riblet drag reduction. *J. Ship Res.* 36 (1992) 55–58.

Hooshmand, D., Youngs, R., Wallace, J.M. and Balint, J.L., An experimental study of changes in the structure of a turbulent boundary layer due to surface geometry changes. *AIAA* Paper 83-0230 (1983).

Kawamura, H., A $k - \varepsilon - \overline{v^2}$ model with special relevance to the near wall turbulence. *Proc. Eighth Symposium on Turbulent Shear Flows* Munich (1991) pp. 26–4–1 to 26–4–6.

Khan, M.M.S., A numerical investigation of the drag reduction by riblet surface. *AIAA* paper 86-1127 (1986).

Kim, J., Study of turbulence structure through numerical simulations: the perspective of drag reduction. *AGARD Report* 786 (Special Course on Skin Friction Drag Reduction) (1992) pp. 7–1 to 7–14.

Klebanoff, P.S., Characteristics of turbulence in a boundary layer with zero pressure gradient. *NACA Report* 1247 (1955).

Lam, C.K.G. and Bremhorst, K.A., Modified form of the $k - \varepsilon$ model for predicting wall turbulence. *J. Fluids Eng.* 103 (1981) 456–460.

Launder, B.E. and Li, S.P., Prediction of drag reduction by riblets. Presentation at the 6th Drag Reduction Meeting, Eindhoven (1991).

Launder, B.E. and Sharma, B.I., Application of the energy-dissipation model of turbulence to the calculation of flow near a spinning disc. *Letters in Heat Mass Transfer* 1 (1974) 131–138.

Launder, B.E. and Tselepidakis, D.P., Contribution to the second-moment modeling of sublayer turbulent transport. In: Kline, S. J. and Afgan, N. H. (eds), *Near-Wall Turbulence*. New York: Hemisphere (1990) pp. 818–833.

Mansour, N.N., Kim, J. and Moin, P., Near-wall $k - \varepsilon$ turbulence modelling. *AIAA Jnl.* 27 (1989) 1068–1073.

Nakamura, I., Miyata, M., Kushoda, T. and Kagiya, K., Some measurements in the intermittent region of a turbulent boundary layer along a corner. In: Fernholz, H. H. and Drause, E. (eds), *Three Dimensional Turbulent Boundary Layers*. Berlin: Springer (1982) pp. 118–209.

Nash, J.F. and Patel, V.C., *Three Dimensional Turbulent Boundary Layers*. SBC Technical Books (1972).

Patankar, S.V., *Numerical Heat Transfer and Fluid Flow*. Series in Computational Methods in Mechanics and Thermal Sciences. McGraw-Hill (1980).

Patel, V.C., Rodi, W. and Scheuerer, G., Turbulence models for near-wall and low-Reynolds number flows: a review. *AIAA Jnl.* 23 (1985) 1308–1319.

Rodi, W. and Mansour, N.N., Low Reynolds number $k - \varepsilon$ modelling with the aid of direct simulation data. In: *Proc. of the Summer Program 1990*. Stanford University: Center for Turbulence Research (1990) pp. 85–106.

Schneider, G.E. and Zedan, M., A modified strongly implicit procedure for the numerical solution of field problems. *Num. Heat Transfer* 4 (1981) 1–19.

Spalart, P.R., Direct simulation of turbulent boundary layer up to $R_\theta = 1410$. *J. Fluid Mech.* 187 (1988) 61–98.

Vukoslavcevic, P., Wallace, J.M. and Balint, J.L., On the mechanism of viscous drag reduction using streamwise aligned riblets: a review with new results. In: *Proc. R.A.S. Int. Conf. on Turbulent Drag Reduction by Passive Means*, Vol. 2. The Royal Aero. Soc. (1987) pp. 290–309.

Walsh, M.J., Turbulent boundary layer drag reduction using riblets. *AIAA* Paper 82-0169 (1982).

Walsh, M.J. and Weinstein, L.M., Drag and heat transfer on surfaces with small longitudinal fins. *AIAA* Paper 78-1161, presented at AIAA 11th Fluid and Plasma Dynamics Conference, Seattle, WA (1978).

Wei, T. and Willmarth, W.W., Reynolds-number effects on the structure of a turbulent channel flow. *J. Fluid Mech.* 204 (1989) 57–95.

Applied Scientific Research **50**: 283–298, 1993.
K. Krishna Prasad (ed.), Further Developments in Turbulence Management.

On the prediction of riblet performance with engineering turbulence models

B.E. LAUNDER & S.P. LI
UMIST, Manchester, U.K.

Abstract. The paper reports the outcome of a numerical study of fully developed flow through a plane channel composed of ribleted surfaces adopting a two-equation turbulence model to describe turbulent mixing. Three families of riblets have been examined: idealized blade-type, V-groove and a novel U-form that, according to computations, achieves a superior performance to that of the commercial V-groove configuration. The maximum drag reduction attained for any particular geometry is broadly in accord with experiment though this optimum occurs for considerably larger riblet heights than measurements indicate. Further explorations bring out a substantial sensitivity in the level of drag reduction to the channel Reynolds number below values of 15 000 as well as to the thickness of the blade riblet. The latter is in accord with the trends of very recent, independent experimental studies.

Possible shortcomings in the model of turbulence are discussed particularly with reference to the absence of any turbulence-driven secondary motions when an isotropic turbulent viscosity is adopted. For illustration, results are obtained for the case where a stress transport turbulence model is adopted above the riblet crests, an elaboration that leads to the formation of a plausible secondary motion sweeping high momentum fluid towards the wall close to the riblet and thereby raising momentum transport.

Nomenclature

- c_f = Skin friction coefficient
- c_{f0} = Skin friction coefficient in smooth channel at the same Reynolds number
- k = Turbulent kinetic energy
- k^+ = $\rho k/\tau_w$
- h = Riblet height
- S = Riblet width
- H = Half height of channel
- Re = Reynolds number = volume flow/unit width/ν
- $\tilde{R}_t$ = Modified turbulent Reynolds number $k^2/\nu\tilde{\varepsilon}$
- R_t = turbulent Reynolds number $k^2/\nu\varepsilon$
- P_k = Shear production rate of k, $\nu_t(\partial U_i/\partial x_j + \partial U_j/\partial x_i)\,\partial U_i/\partial x_j$
- dP/dz = Streamwise static pressure gradient
- U_i = Mean velocity vector (tensor notation)
- U_τ = Friction velocity, $\sqrt{\tau_w/\rho}$ where $\tau_w = -H\,dP/dz$
- W = Mean velocity
- W_b = Bulk mean velocity through channel
- y^+ = yU_τ/ν. Unless otherwise stated, origin is at wall on trough plane of symmetry
- ν = Kinematic viscosity
- ν_t = Turbulent kinematic viscosity
- ε = Turbulence energy dissipation rate
- $\tilde{\varepsilon}$ = Modified dissipation rate $\varepsilon - 2\nu(\partial k^{1/2}/\partial x_j)^2$
- ρ = Density
- σ_k, σ_ε = Effective turbulent Prandtl numbers for diffusion of k and ε

1. Introduction

While the fact that surfaces covered with microgroove riblets can achieve drag levels that are some 5–10% below that of an aerodynamically smooth surface is beyond doubt [1–5], the mechanism responsible for that drag reduction is still open to debate.

Without a clear understanding of the mechanism, it is difficult to arrive by experiment at an optimum design of riblet profile. Inevitable uncertainties in the measurement of skin friction and in embedding a patch of ribleted surface in the wind tunnel usually make it difficult to distinguish a one or two percent advantage from one riblet shape over another from the background scatter.

Against this background the idea of using CFD to guide the optimization of riblet surfaces appears an attractive option. With care in setting up the grid one can, for all practical purposes, eliminate the "noise" – in this case numerical errors – that clouds experimental work. Moreover, with a surface-conforming solver it is a straightforward matter to scan through different families of riblet profiles in search of the optimum. The task, of course, requires considerable computer resource but the overall cost of a numerical investigation is still but a small fraction of that of the corresponding experiments – particularly as so much more detail of the flow is computed than can possibly be measured in an experiment.

The above discussion has not, however, touched on the Achilles heel of computational approaches to riblet modelling – the turbulence model. There is inevitably serious uncertainty as to whether available engineering models of turbulence capture enough of the physics of the near-wall sublayer to allow their extrapolation to flows which exhibit two directions of rapid change within the turbulent buffer region. Previous attempts at modelling riblet flows [6, 7, 8, 21] have employed mixing-length or one-equation models. These can never provide a way of *predicting* drag reduction, for arbitrary choices have to be made – or at least choices unrelated to the flow's physics – on how to interpret, in the vicinity of the riblet, the effective distance of a point from the surface and the appropriate value of wall friction in defining the sublayer velocity scale based on wall friction. Another weakness so far as earlier attempts are concerned has been inadequate numerical resolution in the region near the tip of a riblet. For example, Khan [6] employed ten grid nodes distributed equally along each face of the V-groove riblets considered in his study. Our calculations suggest, however, that the top 10% of his riblets covered by a single node in fact carry up to 40% of the drag of the whole plate!

The simplest level of modelling at which *ad hoc* decisions about wall distance can be avoided is the two-equation level; for then, with transport equations solved for two turbulence variables, effective time and – more importantly – length scales are predicted as an output. Even at two-equation level several of the available models *do* introduce non-dimensional forms of distance from the wall into damping functions designed to make the turbulent kinetic energy and shear stress fall to zero sufficiently rapidly as the wall is approached [9, 10]. However, the original $k-\tilde{\varepsilon}$ model of [11] (or the variant of it [12] which is the principal model adopted in the present study) makes no appeal to wall distance or to the topography of the surface: "proximity" to the wall is sensed only through the local value of the turbulent Reynolds number $\tilde{R}_t \equiv k^2/\nu\tilde{\varepsilon}$ (symbols are defined under Nomenclature). The model has been successfully applied to predict laminarization by acceleration [13] and, more recently, the transition from laminar to turbulent flow triggered by high levels of external stream turbulence [14].

It thus seemed the most appropriate candidate model to employ for this exploration of riblet flows. The aim of the exploration has been to apply the model first to riblets of simple shape to determine whether drag reductions in line with experiment were achieved. A first contribution to this initial study has been reported in [15] but here that work is extended to include blade riblets of finite thickness. This step helps clarify what had seemed an inexplicably large difference in performance between the two riblet types. Being sufficiently encouraged by aspects of these predictions (albeit acknowledging that weaknesses remained), further explorations have been made to see whether some other riblet shape might lead to greater drag reduction than those currently adopted.

One inherent weakness of adopting an eddy viscosity model of turbulent transport in riblet flows is that they are incapable of predicting secondary motions in the plane of the riblet cross section. While no such secondary motions have been reported from experimental studies, it seems probable – based on the wider experience of flow in non-circular ducts – that they do exist and that their presence worsens the performance of the riblet as a drag reducer. Khan's [6] early contribution did include the algebraic model of Launder and Ying [16] to approximate the anisotropy of the stress field acting in the plane orthogonal to the primary flow and, as a result, generated a secondary motion. However, the proposal of [16] was designed for fully turbulent regions only and would be likely to give unreliable results when applied in the buffer layer. Whether for that reason or another the secondary flow eddies generated in Khan's study seemed physically implausible, extending some six riblet heights above the surface and causing, according to the author, a *reduction* in drag.*

A more general route to allowing secondary motions to appear would be through a complete second-moment closure. No such computations have so far been attempted for riblet flows, however. A major obstacle to attempting such an exploration is that all the available models of this type presume that the rigid surface can be regarded as flat – thus explicitly invalidating their use in the vicinity of the riblet itself. However, in order to indicate the type of change to the predicted behaviour that a second-moment closure (when one becomes available) might bring, the present contribution reports sample computations in which a second-moment closure has been adopted for the region above the riblet while retaining the $k-\varepsilon$ model of [12] within the riblet cavity.

2. The mathematical and computational model

The present explorations have focused on the *fully developed flow between infinite parallel planes* covered with riblets aligned with the flow direction. While the principal interest in riblets arises in developing flows, the present limitations should be of no real consequence for the riblets are so small that their tips usually lie within the

*In a computational scheme the contribution of the secondary motion to the wall friction can be readily determined by repeating the calculation with secondary motion set to zero.

(virtually) constant stress layer whether the flow in question is a boundary layer or channel flow. The choice of an infinite plane channel allows considerable computational economy as the computations are then two-dimensional, being solved on a single plane orthogonal to the mean flow. A further advantage of a channel-flow study is that the results are "cleaner": there is no streamwise variation of wall friction or shear layer thickness as there is in a boundary layer flow. The different contributions of the influential parameters can thus be separately assessed.

Most of the computations reported in the paper have adopted an isotropic viscosity model to approximate the turbulent stresses and, as a result, no secondary motions are generated. The distribution of mean velocity in fully-developed flow is thus governed by the following equation:

$$\frac{1}{\rho}\frac{dP}{dz} = \frac{\partial}{\partial x}\left((\nu+\nu_t)\frac{\partial W}{\partial x}\right) + \frac{\partial}{\partial y}\left((\nu+\nu_t)\frac{\partial W}{\partial y}\right), \tag{1}$$

where z is the stream direction and x and y lie in the cross-sectional plane with the coordinate y normal to the wall. In fact, it has been found that variations of velocity and other flow properties in the x direction only occur within three riblet heights of the channel floor. This feature is exploited in obtaining the numerical solution of equation (1). The turbulent viscosity ν_t given by

$$\nu_t = c_\mu k^2/\tilde{\varepsilon}$$

where, following [12], the turbulence energy k and the modified dissipation $\tilde{\varepsilon}$ are obtained from

$$0 = \frac{\partial}{\partial x_j}\left(\left(\nu + \frac{\nu_t}{\sigma_k}\right)\frac{\partial k}{\partial x_j}\right) + P_k - \tilde{\varepsilon} - 2\nu(\partial k^{1/2}/\partial x_j)^2, \tag{2}$$

$$0 = \frac{\partial}{\partial x_j}\left(\left(\nu + \frac{\nu_t}{\sigma_\varepsilon}\right)\frac{\partial \tilde{\varepsilon}}{\partial x_j}\right) + c_{\varepsilon 1}\frac{P_k\tilde{\varepsilon}}{k} - c_{\varepsilon 2}\frac{\tilde{\varepsilon}^2}{k} + 2\nu\,\nu_t\left(\frac{\partial^2 U_i}{\partial x_j \partial x_k}\right)^2. \tag{3}$$

Indices on the spatial coordinate take the values 1 and 2 corresponding to the x and y directions. Like k, the quantity $\tilde{\varepsilon}$ defined as $\varepsilon - 2\nu(\partial k^{1/2}/\partial x_j)^2$ vanishes at the wall – indeed, it was first introduced [11] specifically to provide a clear wall boundary condition. The coefficients in the model take the standard values shown in Table 2.1.

Equations (1)–(3) have been solved numerically using two versions of the finite-volume elliptic solver TEAM [17, 18], a Cartesian-coordinate version being used for blade or L-shaped riblets and a curvilinear orthogonal version for the other riblet

Table 2.1.

c_μ	$c_{\varepsilon 1}$	$c_{\varepsilon 2}$	σ_k	σ_ε
$0.09\,e^{-3.4/(1+0.02\tilde{R}_t)^2}$	1.44	$1.92(1-e^{-\tilde{R}_t^2})$	1.0	1.3

profiles. Following extensive grid refinement tests, typically 80 nodes were placed between the wall and a distance of two rib heights above the riblet crest and 45 nodes between the symmetry planes passing through the riblet crest and trough. By three riblet heights above the channel floor there are sensibly no lateral variations in any flow property and, accordingly, just a single column of nodes is used to cover the region from there to the symmetry plane of the channel. Examples of the near-riblet meshes adopted for the riblets in the present study appear in Fig. 1. Notice that the curvilinear orthogonal mesh naturally leads to a concentration of nodes near the riblet tip which is precisely where it is needed. Although, as a consequence, the near-wall node for the V-groove riblets for $S/h = 1$ is extremely coarse (covering about 20% of the riblet surface), the surface friction over that cell contributes less than 1% of the total drag.

As signalled in Section 1, illustrative computations have also been made for idealized L-shaped riblets with a second-moment closure adopted over the bulk of the flow wherein transport equations are solved for all six Reynolds stress components. Since such a non-isotropic model of turbulence could be expected to generate secondary motions, the describing equations of motion must now include convective transport terms in the x and y directions and, moreover, in addition to the stress-transport equations, additional equations of motion are solved for the x- and y-directed velocities.

3. The computed behaviour

Practical interest centres on whether or not drag reduction is predicted for a given shape and size of riblet. Accordingly, this question is considered first before details of

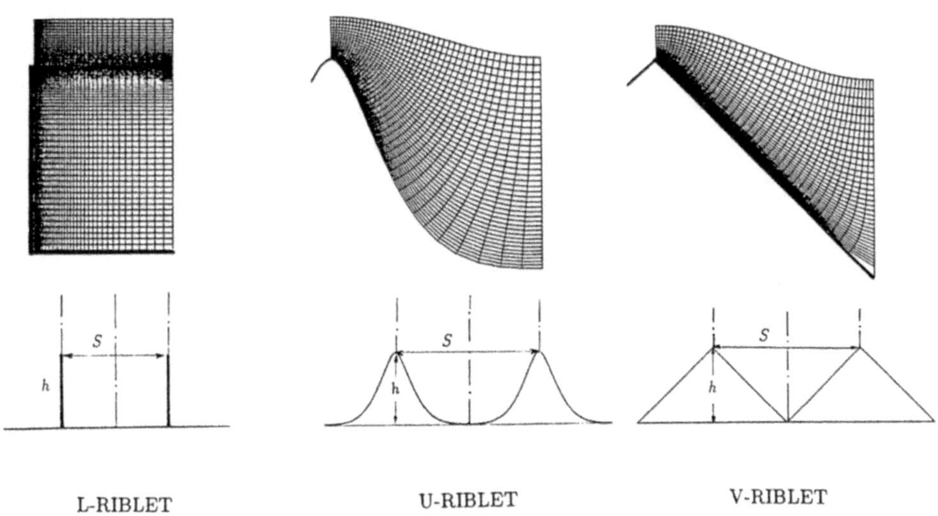

Fig. 1. The riblet configurations studied.

the flow field are examined. Figures 2 and 3 present the computed behaviour for riblet families of idealized L-shaped (with zero thickness) and V-groove forms, respectively. In Fig. 2a the performance at a fixed bulk Reynolds number of 50 000 is shown. Strikingly evident is the fact that appreciable drag reductions *are* predicted over a range of h^+, the optimum performance being achieved for $S/h = \approx 1$, in accord with the conclusions of [3]. However, the magnitude of the drag reduction at $S/h = 1.0$ is at least twice that reported in the experimental literature and occurs at a value of h^+

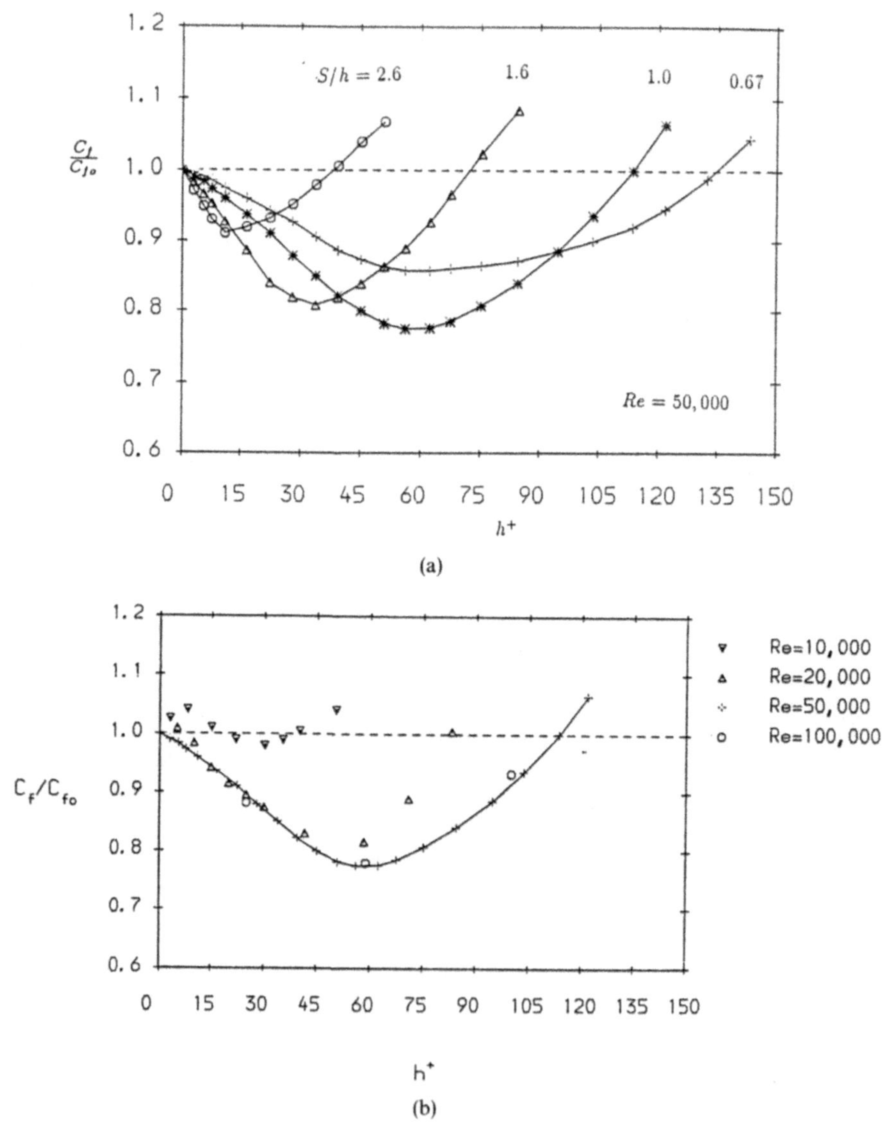

Fig. 2. Predicted drag behaviour for idealized L-shaped riblets. (a) Effect of $S:h$ ratio for $Re = 50\,000$ (from Launder and Li [15]). (b) Effect of Reynolds number on drag reduction for $S/h = 1.0$.

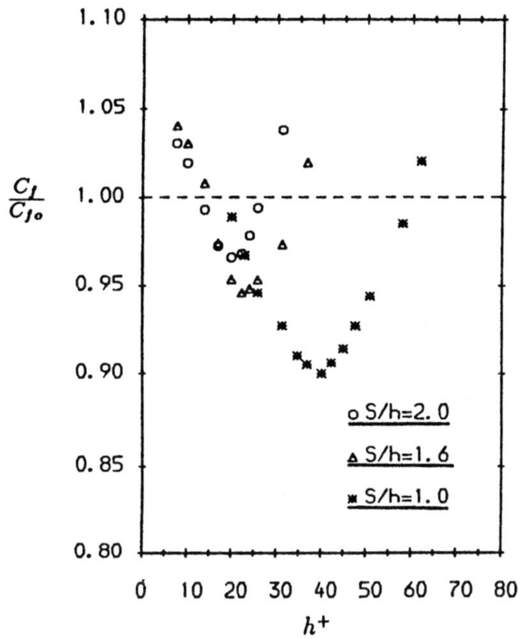

Fig. 3. Drag reduction behaviour for V-grooved riblets.

more than four times that giving the optimum drag reduction in experiments. Both these matters will be examined shortly. First we note, however, from Fig. 2b that, while riblet performance is virtually independent of Reynolds number above $Re = 20\,000$, there is a precipitate decline below that level due partly to the fact that, at a given h^+, the riblet height increases steadily relative to the channel width and thus the velocity near the riblet tip progressively increases relative to $\bar{W}$. Another contributory factor is that in a smooth channel the present turbulence model gives a thickening of the viscous sublayer below $Re = 15\,000$ (a behaviour in accord with channel flow data) giving lower values of c_{f_o} than would have pertained had the smooth-channel predictions rigorously followed the "universal" near-wall velocity profile.

The present experimental data base for riblets is insufficiently complete to distinguish unambiguously possible Reynolds number effects* but it may be noted that the recent study of Benhalilou et al. [7] which adopted large riblets to facilitate the measurement of velocity profiles near the surface (thus producing values of h/δ of order 0.1) show negligible levels of drag reduction.

The drag reduction behaviour for V-groove riblets is summarized in Fig. 3 which provides results where h/H has been fixed at 0.05 in all cases. This means that the bulk

*At low Reynolds number, there must in any event be some doubt as to how precisely fully-developed channel flow provides results analogous to those in a boundary layer because in the former case there will be a significant (~5%) drop in total shear stress in the immediate vicinity of the riblets; a more uniform near-wall shear stress will prevail in the boundary layer, however.

Reynolds number decreases as h^+ diminishes. Here a rather different picture emerges. The greatest drag reduction achieved was now only 10% for $S/h = 1.0$ (where $Re \approx 35\,000$) while for $S/h = 1.6$ and 2.0 the corresponding maximum drag reductions were approximately 5.5 and 4% respectively. These values are closely in line with the reported maximum reductions obtained from experiments on this family of riblets. For example, Sawyer and Winter [19] report 4% for $S/h = 2.1$ and 7% for $S/h = 2$. Admittedly, the values of h^+ at which the optimum drag occurs are still larger than in the experiments but not as seriously as with the idealized L-shaped riblets.

In view of the great difference between the predicted drag reductions obtained with L-profile and V-groove riblets, a limited exploration of the effect of finite riblet thickness was made. In fact, at the time the computations were made, the thinnest L-riblet for which measurements had been reported [4], had a thickness t that was 8% of h. Accordingly, a few runs were made with the same height to thickness ratio for $S/h = 1.0$, Fig. 4. One notices in fact a very marked deterioration of performance with the maximum drag reduction of 11% now being only slightly better than for the V-groove elements. In a forthcoming paper Bruse et al. [20] present experimental results that also show a systematic improvement in riblet performance as t/h is reduced, confirming the trend of our results. In their case a maximum drag reduction of about 8% was achieved with $t/h \approx 0.08$, while for $t/h = 0.02$ the peak reduction reached 10%.

While acknowledging that significant differences remained, the results presented so far suggested that the adopted two-equation model predicted broadly the correct trends so far as the attainable levels of drag reduction are concerned. It thus seemed instructive to employ the curvilinear code in a search for a more efficient riblet profile than offered by the conventional V-groove form. Figure 5 presents relative drag curves

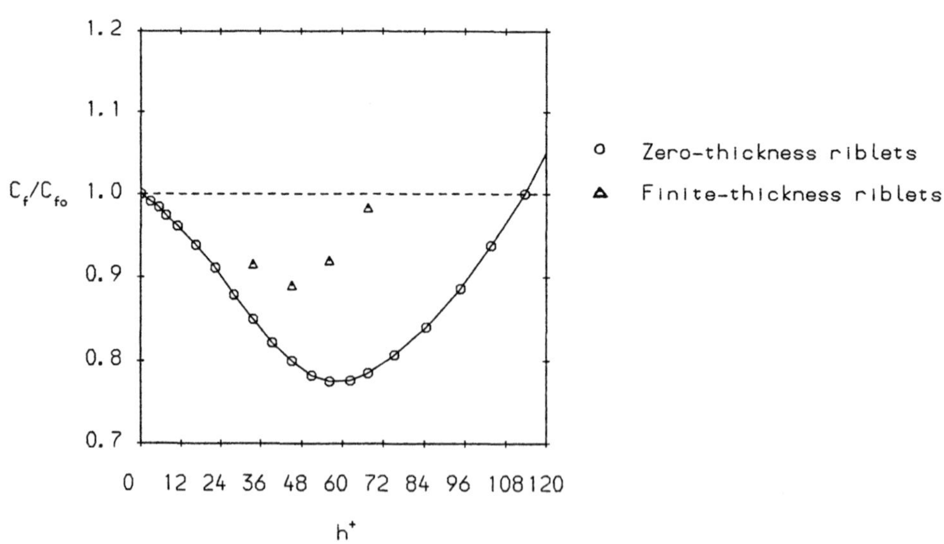

Fig. 4. Effect of finite riblet thickness on drag reduction of blade riblet, $S/h = 1$.

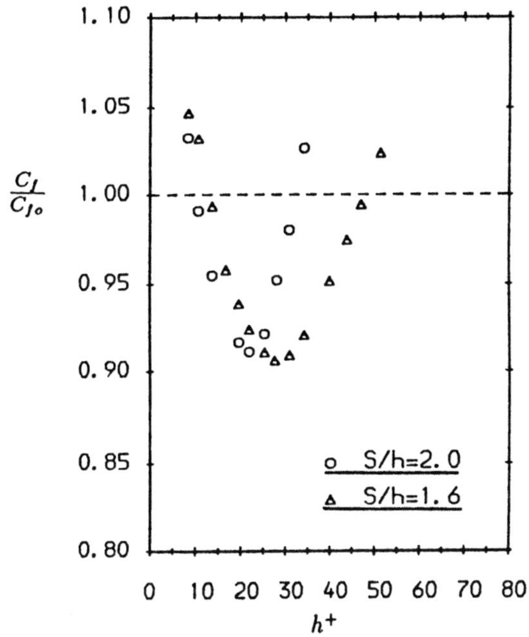

Fig. 5. Drag reduction behaviour for a new family of U-shaped riblets.

obtained for the family of U-shaped riblets shown in Fig. 1. A distinctly better performance is achieved in this case than for the V-groove riblets at S/h of 2.0 and 1.6.*

Figure 6 provides for $S/h = 1.6$ an indication of how the proportion of the total drag on the surface is distributed over the riblet surface for the three riblet profiles. The ordinate represents the fraction of the total drag due to the part of the surface between the trough plane of symmetry and the point in question. Evidently the shear stress is higher at the base of the riblet for the L-riblet than for the other two types since the integrated drag rises more steeply. Due to the very low shear stress levels in the corner, however, the curve for the L-riblet crosses those for the V- and U-riblets. It is striking how much of the drag is carried by the tip region: for L-riblets over 30% of the total drag arises from the 5% of the surface nearest the riblet tip. The frictional loading is distributed rather more evenly in the case of the V-riblet while the U-profile riblet exhibits a distribution intermediate between the other two forms.

Attention is now turned to some of the underlying flow detail giving rise to the behaviour reported in the above figures. Firstly, Fig. 7 presents the near-riblet velocity contours for representatives of the three families of riblets considered. In all cases very steep velocity gradients occur around the crest of the riblet; it is this feature which

*Unfortunately, the termination of the project, coupled with the increased difficulty of securing converged solutions as the width:height ratio decreases, have prevented our obtaining results for $S/h = 1.0$. Nevertheless, the riblet profile shown appears to be one worthy of further study.

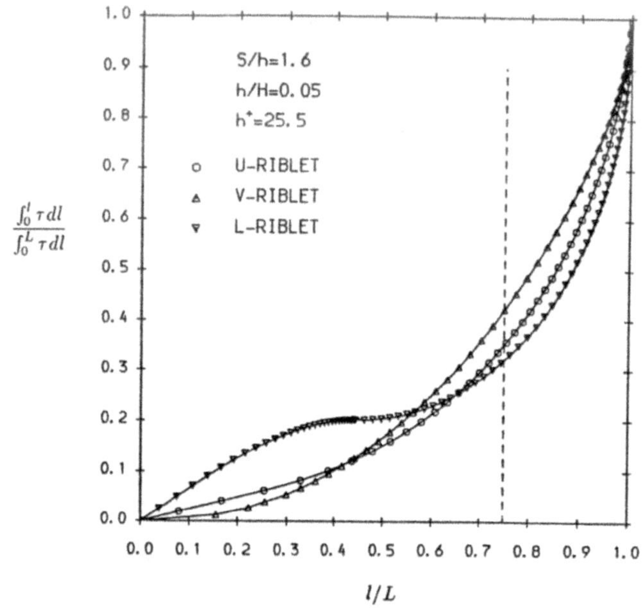

Fig. 6. Integrated contribution of riblet surface to total drag.

makes the performance of the L-profile riblet so sensitive to riblet thickness even though there is a negligible effect on the flow within the cavity itself. Note also that the 0.05 contour runs closer to the riblet surface for the V-groove than for either the U-form or L-profile riblets: this is why it achieves the poorest drag reduction

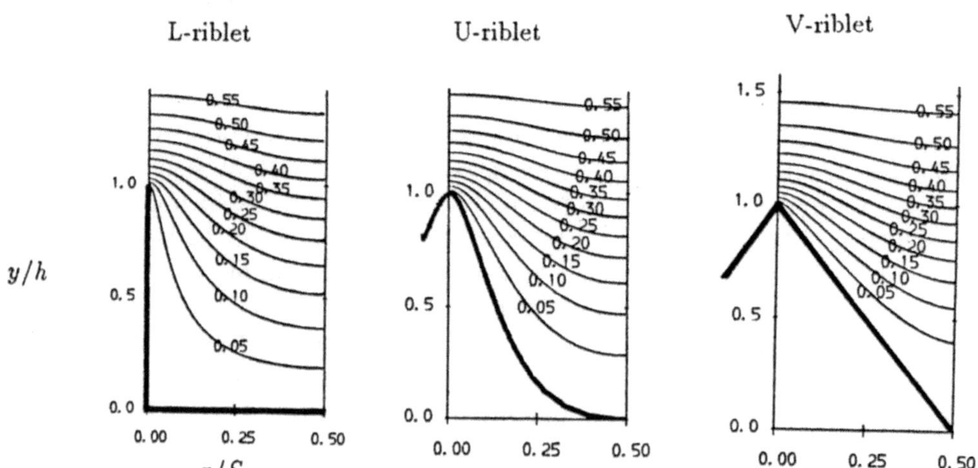

Fig. 7. Contours of streamwise velocity in the vicinity of riblets ($Re=2\times10^4$, $S/h=1.6$, $h/H=0.05$, $h^+\simeq25.5$).

performance of the three despite having the smallest surface area. The corresponding contours of turbulent viscosity appear in Fig. 8a. In all cases the riblets virtually annihilate turbulent mixing within the cavity itself but there are small differences between riblets in the rate at which ν_t increases above the riblet crest, the rate of increase being fastest for the V-groove and slowest for the idealized L-shaped riblets. From Fig. 8b it is seen that, along the trough plane of symmetry, reducing S/h to unity reduces the turbulent viscosity at a given height while enlarging it to 2.0 increases it. These differences in turbulent viscosity stem from modifications to the turbulent kinetic energy profiles near the wall. Figure 9 shows profiles of k^+ for V- and U-profile riblets for two S/h ratios. The predicted distribution arising in smooth channel flow at

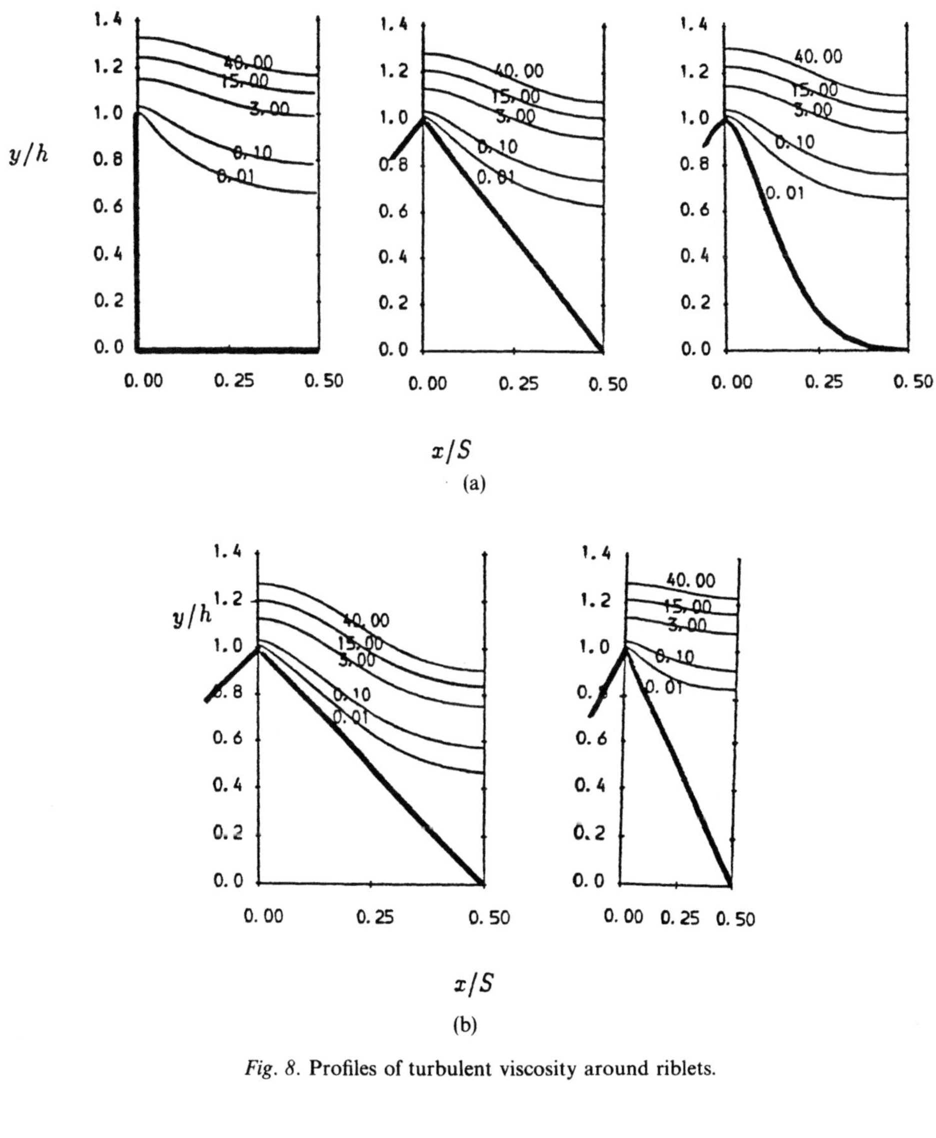

Fig. 8. Profiles of turbulent viscosity around riblets.

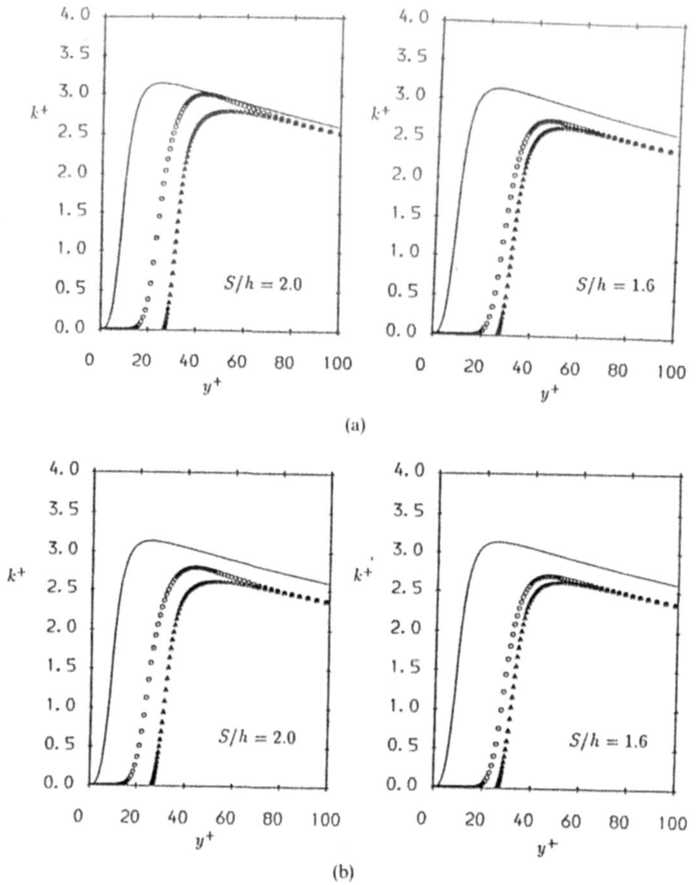

Fig. 9. Computed profiles of normalized turbulence energy, —— smooth channel; ○ riblet trough; Δ riblet crest, (a) V-groove riblets; (b) U-shaped riblets.

the same *Re* with the same turbulence model is also included for reference (the coordinate y^+ has its origin in the trough of the riblet for both crest and trough profiles). For both riblets the steep rise in k^+ across the buffer layer is displaced to larger values of y^+ indicating the thickening of this viscosity affected sublayer. Moreover, the peak levels of k^+ are noticeably less than for a smooth surface, particularly so for the smaller S/h and the U riblets.

4. Discussion

The computed behaviour reported in Section 3 exhibits encouraging and discouraging features. Certainly the possibility of riblets achieving a drag reduction has emerged very clearly from the computations and the different levels of drag reduction achievable for different riblet configurations is broadly in line with the experiment.

However, as noted in [15], two different and counteracting weaknesses in the physical model are plainly present. At small riblet heights ($h^+ < 10$) the present model gives too little drag reduction while at larger values ($h^+ > 15$) the overall performance is greater than found experimentally at the same riblet heights. The former weakness may partly arise from the specific dependency of c_μ on Re_t: a distribution of $c_\mu(Re_t)$ that gives the correct average level of turbulent viscosity across the buffer region on a strictly smooth surface may not do so when riblets intervene. An alternative source of discrepancy may be the $\tilde{\varepsilon}$ equation. It seems plausible that flow inhomogeneities parallel to the (mean) rigid surface (in addition to inhomogeneities normal to the surface) should tend to raise ε, thus decreasing k and the turbulent viscosity. It is noted that through the final term in equation (3) inhomogeneities in the sublayer in x as well as y do tend to raise $\tilde{\varepsilon}$. Perhaps some further source term, involving gradients in k and/or ε ought to be added to the right-hand side of the $\tilde{\varepsilon}$ equation.

The most likely reason that our predictions show strong drag reductions to values of h^+ that are 2–3 times greater than reported in experiments is the absence of secondary flows from our calculations (an inevitable consequence of adopting an isotropic viscosity model). The remedy is to shift instead to a second moment closure. However, while second moment closures are now beginning to be used in engineering flows near walls (see the reviews of [22] and [23]), within the viscosity affected sublayer simpler schemes are invariably adopted, at least if the sublayer is three-dimensional. The use in the model of the unit vector normal *to the wall* or the distance *from the wall* make the available models unsuitable for strongly non-planar surfaces such as riblets. However, to go at least some way in the direction of the proposed model refinement for the case of the idealized blade riblet, a hybrid treatment has been adopted. Within the riblet cavity itself, the two-equation $k-\tilde{\varepsilon}$ model adopted for the main body of our calculations has been retained. At a height of $1.2h$ from the floor, however, a switchover is made to a second-moment closure that has been very successfully applied to the prediction of secondary flow on rectangular ducts (the model is algebraically very complex and is not reproduced here; details are given in [8, 24]). For this computation, as noted at the end of Section 2, the full momentum equations are solved in all three Cartesian directions and, in the stress transport equations, the normal-to-the-wall direction is taken as y. Only within the region where the second-moment closure is adopted does the turbulence model provide a source of streamwise vorticity. For this reason, we may expect the adoption of the hybrid model to generate a smaller secondary flow than would arise if a second-moment closure could have been used throughout. We note from Fig. 10, however, that a distinct secondary eddy is nevertheless present, centred at about one third of the riblet height above the tip, which sweeps down high velocity fluid near the riblet with a more diffuse return flow mid-way between riblets. The secondary motion distorts the axial velocity contours in the sense expected and leads to an increase in momentum transport (and thus of total drag) of about 3%. As remarked above, we may expect that a full second-moment treatment would produce even larger momentum transport rates by the secondary motion.

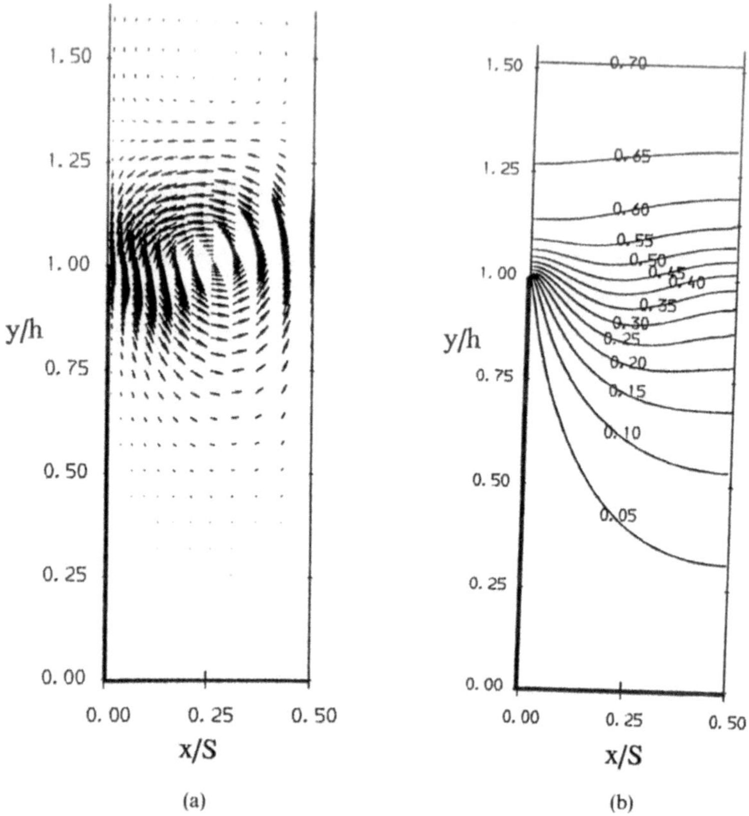

Fig. 10. Predicted flow pattern in L-shaped riblet for $S/h = 1.0$, $h/H = 0.06$ ($h^+ \simeq 67$). (a) Secondary velocity vectors; (b) contours of W/W_{max}.

5. Conclusions

1. A two-equation eddy viscosity model has been applied to predict the flow around L-shaped and V-grooved riblets. The predicted maximum levels of drag reduction, including the effect of height:pitch ratio and, in the case of L-shaped riblets, the riblet thickness, are reasonably in accord with measurements.
2. A substantial sensitivity to Re has emerged for channel flow Reynolds numbers below about 15 000 with the predicted benefits of riblets virtually disappearing below $Re = 10^4$.
3. Further explorations have been made of a family of U-shaped riblets which, according to our predictions, would give better performance than the usual V-grooved form.
4. While the optimum drag reduction levels predicted in our study are broadly correct, there are nevertheless two substantial shortcomings in the computations: below $h^+ \sim 10$ too little drag reduction is achieved while for $h^+ > 15$ too large

reductions are indicated. This contrasting behaviour springs from two counteracting weaknesses in the model: an insufficient sensitivity to x-direction inhomogeneity on the level of ε and the use of an isotropic viscosity formulation that suppresses secondary motion.

5. A limited exploration has been made for the case of an L-shaped riblet with $h/s = 1.0$ in which in the region above the riblets, a second-moment closure has been used in place of the eddy viscosity model. The use of this model produces a streamwise secondary vortex on either side of the riblet that sweeps high momentum fluid down near the riblet surface thus raising drag.

Acknowledgements

The research has been sponsored by the SERC under Grant GR/E 77039. S-P Li also acknowledges with thanks support from the British Council and the Northwestern Polytechnic University, Xian, PRC. During the research Dr A.M. Savill has made numerous appreciated technical inputs to the work programme. Mrs J. Buckley and Mrs D. Lindop have processed the manuscript for publication. Authors' names are sequenced alphabetically.

References

1. Walsh, M.J. and Weinstein, L.M., Drag and heat transfer characteristics of small, longitudinally ribbed surfaces. *AIAA J.* 17 (1979) 770.
2. Walsh, M.J., Turbulent boundary layer drag reduction using riblets. *AIAA* Paper 82-0169 (1982).
3. Wilkinson, S.P., Anders, J.B., Lazos, B.S. and Bushnell, D.M., Turbulent drag reduction research at NASA Langley: progress and plans. *Int. J. Heat Fluid Flow* 9 (1988) 286.
4. Wilkinson, S.P. and Lazos, B.S., Direct drag and hot-wire measurements on thin-element riblet arrays. IUTAM Symposium on Turbulence Management and Relaminarization, Bangalore (1987).
5. Walsh, M.J., Optimization and application of riblets for turbulent drag reduction. *AIAA* Paper 84-0347 (1984).
6. Khan, M.M.S., A numerical investigation of the drag reduction by riblet surfaces. *AIAA* Paper 86-1127 (1986).
7. Benhalilou, M., Anselmet, F., Liandrat, J. and Fulachier, L., Experiments and numerical investigation of a turbulent boundary layer over riblets. *Proc. 8th Symp. Turbulent Shear Flows*, Paper 18-5, Munich (1991).
8. Li, S.-P., PhD Thesis, Faculty of Technology, University of Manchester (1992).
9. Chien, K.-Y., Predictions of channel and boundary-layer flows with a low-Reynolds-number turbulence model. *AIAA Journal* 20 (1982) 33.
10. Lam, C.K.G. and Bremhorst, K.A., Modified form of the $k-\varepsilon$ model for predicting wall turbulence. *J. Fluids Engineering* 103 (1981) 456.
11. Jones, W.P. and Launder, B.E., The prediction of laminarization with a two-equation model of turbulence. *Int. J. Heat Mass Transfer* 15 (1972) 301.
12. Launder, B.E. and Sharma, B.I., Application of the energy-dissipation model of turbulence to the calculation of flow near a spinning disc. *Letters in Heat and Mass Transfer* 1 (1974) 131.
13. Patel, V.C., Rodi, W. and Scheuerer, G., *AIAA Journal* 23 (1985) 1320.
14. Savill, A.M., Turbulence model predictions for transition under free stream turbulence, Unpublished report, Engineering Department, University of Cambridge, 1991. (Also poster paper for Royal Aeronautical Society, Transition and Boundary Layer Conference, Cambridge, April 1991).

15. Launder, B.E. and Li, S.-P., The prediction of riblet behaviour with a low-Reynolds number $k-\varepsilon$ model. *Aeronautical Journal* 96 (1992) 354–358.
16. Launder, B.E. and Ying, W.-M., Prediction of flow and heat transfer in ducts of square cross-section. *Proc. Institution of Mechanical Engineers* 187 455 (1973) 37/73.
17. Huang, P.G. and Leschziner, M.A., An introduction and guide to the computer code TEAM. UMIST Mech. Eng. Dept. Report TFD/83/9 (R) (1983).
18. Leschziner, M.A., Kadja, M. and Lea, C., A combined experimental and computational study of a separated flow in an expanding annular passage. In: Iwasa, Y., Tamai, N. and Wada, A. (eds), *Refined Flow Modelling and Turbulence Measurements.* Tokyo: Universal Academic Press (1988) p. 129.
19. Sawyer, W.G. and Winter, K.G., An investigation of the effect on turbulent skin friction of surfaces with streamwise grooves. *Proc. Symposium Turbulent Drag Reduction by Passive Means.* London: Royal Aeronautical Society (1987).
20. Bruse, D.W., Bechert, D.W., van der Hoeven, J.G.Th., Hage, W. and Hoppe, G., Experiments with conventional and with novel adjustable drag-reducing surfaces. In: So, R.M. et al. (eds), *Proc. Int. Symp. on Near Wall Turbulence.* Elsevier (1993) pp. 719–738.
21. Djenidi, L., Squire, L.C. and Savill, A.M., High-resolution conformal mesh computations for V-, U- or L-groove riblets in laminar and turbulent boundary layers. In: Choi, K.-S. (ed.), *Recent Developments in Turbulence Management.* Dordrecht: Kluwer Academic Publishers (1991) pp. 65–92.
22. Launder, B.E., Turbulence modelling for the nineties: second-moment closure... and beyond? In: Morton, K.W. (ed.), *12th International Conference on Numerical Methods in Fluid Dynamics.* Lecture Notes in Physics 371. Berlin: Springer Verlag (1990) pp. 1–18.
23. Leschziner, M.A., *Int. J. Heat Fluid Flow* 10 (1989) 186–202.
24. Launder, B.E. and Li, S.-P., Contribution to the modelling of turbulent flow in rectangular passages. Submitted for publication (1992).

Applied Scientific Research **50**: 299–314, 1993.
K. Krishna Prasad (ed.), Further Developments in Turbulence Management.

Modelling the time dependent flow over riblets in the viscous wall region

S. TULLIS & A. POLLARD*
Department of Mechanical Engineering, Queen's University at Kingston, Ontario, K7L 3N6 Canada
*(*author for correspondence)*

Abstract. The flow over riblets is examined computationally using a time dependent model of the viscous wall region. This "2 1/2 D model", developed by Hatziavramidis and Hanratty (1979) and modified by Nikolaides (1984) and Chapman and Kuhn (1981, 1986) assumes homogeneity in the streamwise direction so that the flow is solved only in the cross-sectional plane. The flow at the upper boundary of the computational domain ($y^+ \simeq 40$) is described using a streamwise eddy model consisting of two scales, one of the streak spacing ($\lambda^+ \simeq 100$), which dominates vertical momentum transport, and a larger scale that accounts for the influence of large outer flow eddies.

The protrusion height concept (Bechert and Bartenwerfer, 1989) is used to define a $y^+ = 0$ location for surfaces with riblets. A control volume finite element method utilizing triangular meshes is used to exactly fit the riblet cross-sectional geometry. Results obtained using fairly large riblets compare well with the limited experimental evidence available. Observations of the transient flow suggest that the riblets interact with the near-wall streamwise vortices, weakening them by the generation of intermittent secondary vortices within the riblet valleys. The riblets also appear to limit the lateral spread of inrushes towards the wall and retain low momentum fluid in the riblet valleys effectively isolating much of the wall from such inrushes.

1. Introduction

The ability of riblets to produce significant skin friction reductions in turbulent boundary layer flows is well established. Wilkinson et al. (1987), Walsh (1990) and Savill (1989) have suggested that the viscous effect of the riblets in accumulating slow flow with low shear stress in the riblet grooves plays a major role in the skin friction reduction. The effect of riblets, however, has been observed by Vukoslavĉević et al. (1987) and Choi (1987, 1989) to extend well above the viscous wall region. This has led these authors to the conclusion that at least a portion of the drag reduction is a result of the influence of riblets on near-wall turbulent coherent structures.

In the very near-wall region ($y^+ \leqslant 30$–40), the dominant turbulent structures appear to be quasi-periodic streamwise eddies Bakewell and Lumley (1967), Townsend (1956). These eddies act as a mechanism for the transport of streamwise momentum to and away from the wall which then contribute to the high Reynolds stress and turbulence production close to the wall. The near-wall low-speed streaks, which have a characteristic spanwise "streak" spacing of $\lambda^+ \simeq 100$, are generally accepted (Smith et al. 1989) to be the lift up of low momentum fluid from the wall by an adjacent streamwise eddy. The surface drag reduction effect of riblets then seems to involve some sort of influence on these near-wall eddies. An increase in the streak spacing and decrease in spanwise motions over riblet surfaces has been noted experimentally by many researchers, including Gallager and Thomas (1984), Bacher and Smith (1985) and Choi (1989). Both of these effects are similar to the drag reduction produced by dilute polymer solutions (Tiederman, 1989).

A number of models have been proposed which depict how, by interacting with the near-wall streamwise vortices, riblets may reduce the turbulent skin friction. Bacher and Smith (1985) have suggested that the generation of secondary vortices within the riblet valleys weakens these streamwise vortices. Flow visualization work by Clark (1989) appears to show such secondary groove vortices. The development of spanwise concentrations of low-speed fluid (streaks) is then inhibited, leading to streaks with larger spacings and weaker velocity gradients. Smith et al. (1989) proposed that riblets interfere with the lateral (spanwise) motion of low speed fluid (streaks) at the wall. Choi (1989) also suggested that the major effect of the riblets is to restrict the spanwise motion of the streamwise vortices and low speed streaks. The models seem to agree that the riblets probably interfere with both the intensities and the spanwise motions of the near-wall vortices.

Until very recently, the modelling of the flow over riblets has usually been performed with simple half or one riblet geometries and with either laminar flow or mixing-length Djenidi et al. (1990), Benhalilou et al. (1991) and, more recently, $k-\varepsilon$ turbulence models, Djenidi (1991), Launder and Li (1991). These approaches have used Reynolds averaged and often parabolized equation sets to produce steady-state solutions. These methods do not consider multiple riblet interactions with any specific near-wall turbulent structures, particularly streamwise vortices, and depend critically on the near-wall damping assumptions.

The approach taken here is to use a "simple eddy" or "coherent structure" model to attempt to directly model quasi-periodic streamwise eddies in the very near-wall region. The model was originally developed, for flat wall turbulent boundary layers, by Hatziavramidis and Hanratty (1979) and developed and refined by Chapman and Kuhn (1981, 1986) and Nikolaides (1984). The model is not intended to exactly predict the flow; it is a representation of near-wall streamwise eddies and their time development – although the models do provide good agreement with experimentally measured mean velocity, Reynolds stress and turbulence intensity profiles for flat walls.

2. Turbulent boundary layer modelling

2.1. Governing equations

The model of the turbulent boundary layer is time dependent with the coordinates oriented such that the downstream direction is along the x axis, the y direction is perpendicular to the wall and the z axis is across the flow in the spanwise direction. With the streamwise eddies being highly elongated (Kline et al. (1967), Robinson et al. (1989), a major assumption is made that the streamwise velocity derivatives can be neglected in comparison to the cross-stream velocity derivatives. The governing equations are then

$$\frac{\partial v^+}{\partial t^+} + \frac{\partial v^{+2}}{\partial y^+} + \frac{\partial w^+ v^+}{\partial z^+} = -\frac{\partial p^+}{\partial y^+} + \frac{\partial^2 v^+}{\partial y^{+2}} + \frac{\partial^2 v^+}{\partial z^{+2}}, \tag{1}$$

$$\frac{\partial w^+}{\partial t^+} + \frac{\partial v^+ w^+}{\partial y^+} + \frac{\partial w^{+2}}{\partial z^+} = -\frac{\partial p^+}{\partial z^+} + \frac{\partial^2 w^+}{\partial y^{+2}} + \frac{\partial^2 w^+}{\partial z^{+2}}, \tag{2}$$

$$\frac{\partial v^+}{\partial y^+} + \frac{\partial w^+}{\partial z^+} = 0 \tag{3}$$

and,

$$\frac{\partial U^+}{\partial t^+} + \frac{\partial v^+ U^+}{\partial y^+} + \frac{\partial w^+ U^+}{\partial z^+} = \frac{\partial^2 U^+}{\partial y^{+2}} + \frac{\partial^2 U^+}{\partial z^{+2}}. \tag{4}$$

All of the quantities in Eqs 1–4 are normalized by the friction velocity, u_τ, with $u_\tau = \sqrt{\tau_w/\rho}$, where τ_w is the flat wall shear stress. Lengths are non-dimensionalized as

$$x^+ = \frac{u_\tau}{\nu} x \qquad y^+ = \frac{u_\tau}{\nu} y \qquad z^+ = \frac{u_\tau}{\nu} z,$$

where $\nu = \mu/\rho$ is the kinematic viscosity. The velocity components, pressure and time can all be non-dimensionalized as:

$$U^+ = \frac{U}{u_\tau} \qquad v^+ = \frac{v}{u_\tau} \qquad w^+ = \frac{w}{u_\tau},$$

$$p^+ = \frac{p}{\tau_w} = \frac{p}{\rho u_\tau^2} \qquad t^+ = \frac{u_\tau^2}{\nu} t.$$

Wall normalized units will be used for the remainder of this paper, so the superscript "+" will be dropped.

It can be noted that the cross-stream velocities v and w and the down-stream velocity U are not coupled – the cross-stream velocities can be solved using Eqs 1–3, then U can be solved directly using Eq. 4. There is also no downstream pressure gradient.

2.2. Boundary conditions

The computational domain consists of a cross-sectional $y-z$ plane with $0 \leqslant y \leqslant y_0$ where y_0 is typically 40. The spanwise size of the domain depends on the particular variation of the model but is usually in the range of 100–400 wall units.

The main modelling assumption of the method is in the specification of the velocities at the upper edge, $y_0 \simeq 40$, of the domain. These specifications can vary depending on the specific model; however, in all of the models, the velocity components specified at the upper boundary are periodic in time and spanwise location (z) and have overall magnitudes of the experimentally determined turbulence intensities at that distance from the wall. The downstream velocity U is specified to consist of two components: a fluctuating component u and a constant component $\bar{U}$ which is determined, for flat walls, from experimental measurement or from the log-law relationship.

The fluctuating velocity components usually have two scales: one that dominates vertical momentum transport and roughly the size of the streak spacing and a larger scale accounting for the influence of all large "outer flow eddies". The full form as given by Nikolaides (1984) is:

$$U = \bar{U} + \hat{u}_1 \cos\left(\frac{2\pi}{T_1} t + \phi_{u1}\right) \cos \frac{2\pi}{\lambda_1} z + \hat{u}_2 \cos\left(\frac{2\pi}{T_2} t + \phi_{u2}\right) \cos \frac{2\pi}{\lambda_2} z, \tag{5}$$

$$v = \hat{v}_1 \cos\left(\frac{2\pi}{T_1} t + \phi_{v1}\right) \cos \frac{2\pi}{\lambda_1} z + \hat{v}_2 \cos\left(\frac{2\pi}{T_2} t + \phi_{v2}\right) \cos \frac{2\pi}{\lambda_2} z, \tag{6}$$

$$w = \hat{w}_1 \cos\left(\frac{2\pi}{T_1} t + \phi_{w1}\right) \sin \frac{2\pi}{\lambda_1} z + \hat{w}_2 \cos\left(\frac{2\pi}{T_2} t + \phi_{w2}\right) \sin \frac{2\pi}{\lambda_2} z. \tag{7}$$

The periods in the spanwise direction, λ_1 and λ_2, are typically set respectively as 100, the size of the flat wall streak spacing, and 400 or infinity as a larger period accounting for the influence of all larger scale eddies. The time periods T_1 and T_2 are usually set to roughly the same values as λ_1 and λ_2 to agree with bursting frequencies for it is reasoned that the larger scale transient and convective terms should be of the same order, Nikolaides (1984). Another important set of parameters are the amplitudes of fluctuations at the different scales, which are related to the energy fractions of the velocity components in each of the scales. These are selected using mainly experimental data. The phase angles (ϕ's) between the various periodic terms are generally selected based on the experimental measurements of the average size of the near-wall eddies and the use of computer optimization. The model and the values of the constants used in this study are from Chapman and Kuhn (1986). The parameters used are summarized in Table 1.

This model consists of Nikolaides' (1984) model with modifications which, for the

Table 1. Parameters used in Chapman and Kuhn's (1986) model

λ_1	100	$\lambda_2(u, v, w)$	∞, –, ∞
T_1	143	$T_2(u, v, w)$	143, –, 286
u_1	1.8	u_2	2.53
v_1	1.0	v_2	0
w_1	2.0	w_2	1.175
ϕ_{u1}	0 (reference)	ϕ_{u2}	0
ϕ_{v1}	0	ϕ_{v2}	–
ϕ_{w1}	$\pm\frac{\pi}{2}$	ϕ_{w2}	$\frac{2\pi}{3}$

most part, are simplifications that eliminate some of the seemingly arbitrarily selected terms while retaining good agreement with experimental data.

The sides of the computational domain (at $z=0$ and $z=z_0$) are treated with cyclic boundary conditions. The wall surface is specified using the no-slip condition: $U=v=w=0$. The specification of the upper $y_0=40$ conditions requires a definition of the location of the $y=0$ surface. This surface location is simply the wall surface; however, for riblet walls, the issue is not so straightforward. Ignoring the fluctuating components from this viscous wall layer model (Eqs 1–4), there remains:

$$\frac{\partial^2 U}{\partial y^2} + \frac{\partial^2 U}{\partial z^2} = 0. \tag{8}$$

Using conformal mapping for various riblet geometries, Bechert and Bartenwerfer (1989) calculated a protrusion height of the riblet peaks above an effective origin of the U velocity profile, thereby providing a definition of the $y=0$ location for a riblet surface.

2.3. Numerical scheme

The numerical solution procedure used in this study is a Control Volume Finite Element Method (CVFEM) based on the work of Saabas (1991). The approach is an equal order co-located control volume method and is a development of earlier work by Baliga and Patankar (1979), (1983) and Prakash and Patankar (1985). The computational domain is divided into three-node triangular elements that are able to fit exactly such complex geometries as riblets in cross-section. Control volumes are constructed about each node with the complete discretization equations assembled for each control volume on an element-by-element basis. Exponential flow-oriented interpolation functions are used within each element to evaluate the velocity components, which correctly simulate the nature of convection, resulting in a substantial reduction of false diffusion, Baliga (1979). A time-stepping procedure was incorporated into the CVFEM using an implicit scheme, see, for example, Patankar

(1980). A structured grid is used with the triangular elements constructed by connecting the diagonals of a rectangular grid. This allows the use of line-by-line solvers. Space precludes the presentation of more detail.

3. Presentation and discussion of results

Three different cases are considered: flow over a flat wall and flow over both V-groove and U-groove riblet walls. The V-groove and U-groove riblets have dimensions of $h^+=16.7$ and $s=2h$; thus, the computational domain consists of three complete riblets. Riblets with such dimensions neither reduce nor increase drag, Walsh (1990). However, these riblets correspond well with the experimentally investigated cases of Vukoslavĉević et al. (1987) and Benhalilou et al. (1991).

For the flat wall case, a uniform 21×43 (y and z directions) grid is used. For the V-groove and U-groove riblet cases, the number of grid points in the spanwise (z) direction is the same as the flat wall case while additional grid points are used in the y direction – principally to account for the increased domain size resulting from the area between the base of riblet valleys and the protrusion height origin. Since the shape of the riblets depends on the grid spacing, a non-uniform grid is used to produce the semi-circular U-groove riblets. Typically, the grid spacing in the vicinity of the riblet peaks is $\Delta x^+ \simeq \Delta y^+ \simeq 2$, see Tullis (1992).

The solutions consist of approximately 500 time steps with a time step size of $\Delta t^+ = 2$. This time step size was selected from a comparison of flat wall solutions run with the particular spatial grid sizes and various time step sizes. The initial solution within the computational domain at $t = 0$ is the solution to Eq. (4) for the downstream velocity with no cross-stream velocities ($v = w = 0$). The effects of using this simple initial solution are not significant after two of the larger scale time periods (T_2). All space-time averaged quantities are calculated starting well after this initial interval and for over at least two of the larger scale time periods.

3.1. Mean velocity profiles

The model was tested against data obtained on a flat wall turbulent boundary layer; the data of Laufer (1954) and Ueda and Hinze (1975) were used. The comparison is shown in Fig. 1. The agreement is deemed acceptable. There is, however, a noticeable discrepancy near the outer edge of the domain. This effect can be also noticed in many of the other computational results, particularly the turbulent statistics. This irregularity has been observed by Chapman and Kuhn (1986) and Nikolaides (1984) who have suggested that it is the result of too restrictive boundary conditions, probably on the smaller scale eddies, and both consider that this effect is limited to the outer 10–20% of the domain.

The calculated mean streamwise velocity profiles over the V-groove riblets seem to

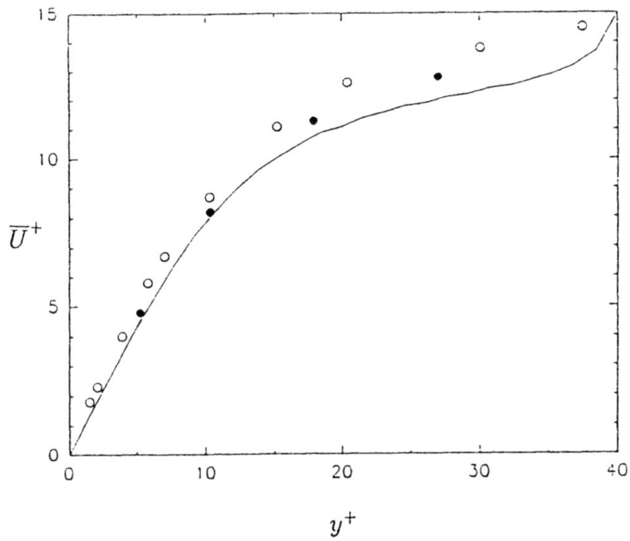

Fig. 1. Mean streamwise velocity ($\bar{U}$) profiles over a flat wall. Solid line: viscous wall region model. (●): data of Laufer, 1954. (○): data of Ueda and Hinze, 1975.

agree well with the experimental measurements of Vukoslavĉević et al. (1987), as shown in Fig. 2. The velocity gradient distribution at the riblet surface shows the same features observed by Vukoslavĉević et al.

As depicted in Fig. 3a, the wall shear stress drops nearly to zero in the angled valleys of the V-groove riblets. Note that s^+ represents distance along the riblet surface. The

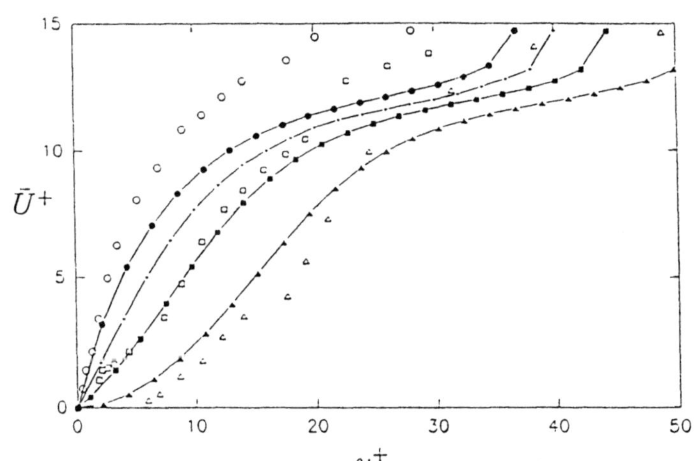

Fig. 2. Mean streamwise velocity profiles over V-groove riblets. Comparison of the calculations (●, ■, ▲) and experimental results (○, □, △) from Vukoslavĉević et al. (1987) for the riblet peaks, mid-point between peaks and valleys, and valleys respectively. The calculated flat wall velocity profile is also shown (+).

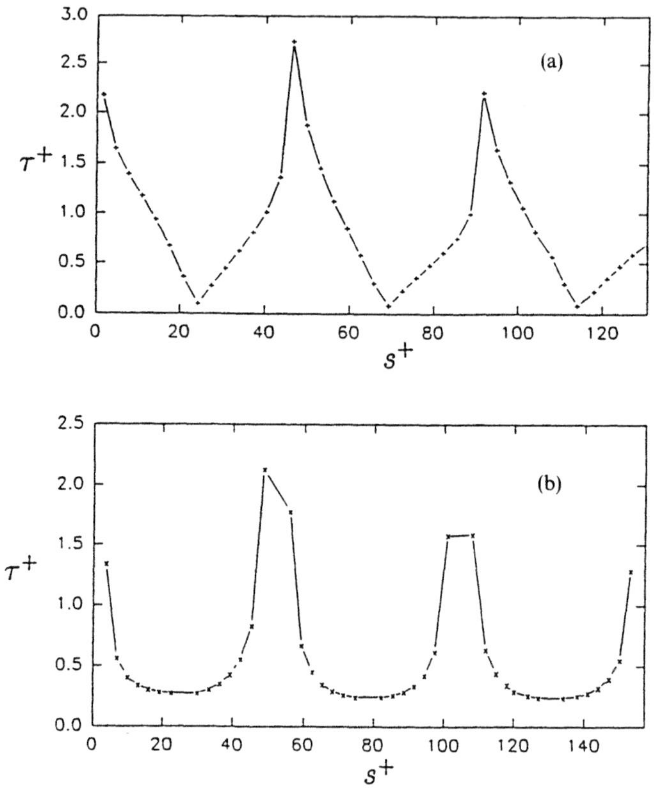

Fig. 3. Calculated dimensionless shear stress ($\tau^+ = \tau/\tau_w$) spanwise distributions for: (a) V-groove riblet wall and (b) U-groove riblet wall. s^+ is distance along riblet wall surface.

wall shear stress over the riblet peaks is much larger than that found over the flat wall – the increase is approximately 150%. This increase is larger than that noted by both Vukoslavĉević et al. and Benhalilou et al. (1991) but may be exaggerated by the relatively coarse grid that is being used in the present calculations. A similar surface drag distribution is observed over the U-groove riblets (Fig. 3b). The wall shear stress in the rounded riblet valleys drops to about 25% of the flat wall case while the increase above the riblet peaks is approximately the same size as that found for the V-groove riblets.

3.2. *Streamwise turbulence intensity*

The calculated turbulence intensities over a flat wall show fairly good agreement with experimental data, not shown for brevity, as was demonstrated by Chapman and Kuhn (1986). Experimental measurements of turbulence intensities within and above riblets are very limited. Vukoslavĉević et al. have measured streamwise velocity fluctuation intensity (u') profiles while Benhalilou et al. were also able to obtain spanwise velocity fluctuation intensity (w') profiles with similarly large-sized V-groove riblets. Consequently, attention will be focussed on the streamwise turbulence intensity profiles. It should be noted that the two sets of experimental data have significant differences. With increasing distance from the wall, Benhalilou et al. observed the u' intensity to increase rapidly over the riblet peaks followed by nearly as rapid decrease, as can be seen in Fig. 4. The intensity over the riblet valley reaches a much lower maximum value and the peak and valley intensities are converging to a flat wall value of approximately 1.5, significantly lower than most other flat wall experimental values such as those of Clark (1968) and Ueda and Hinze (1975), at a height of $y^+ = 35$. Vukoslavĉević et al., in contrast, observe very little decrease in the u' intensity profiles after they reach their maxima – with the peak and valley profiles converging at a higher than flat wall value of $u'^+ \sim 2.4$. The calculated u' intensity profiles resemble those of Vukoslavĉević et al. close to the wall, although the profiles in the valley appear to reach slightly larger values closer to the wall. In the upper half of the calculation domain, however, the peak and valley profiles converge while steadily decreasing to a value of approximately 1.8 at the outer edge of the domain. In this respect, the calculated results are roughly between the two sets of experimental data.

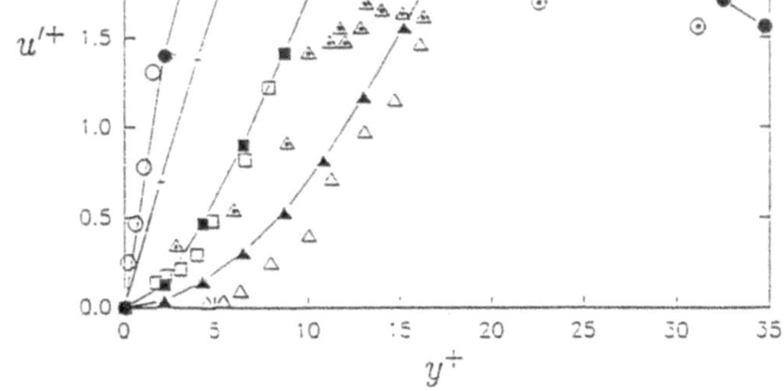

Fig. 4. Streamwise turbulence intensity profiles over V-groove riblets. Comparison of the calculations (●, ■, ▲) and experimental results from Vukoslavĉević et al. (○, □, △) and Benhalilou et al. (⊙, ⧍) for the riblet peaks, mid-point between peaks and valleys, and valleys respectively. Data from Benhalilou et al. do not include results for the profile above the mid-point between peaks and valleys.

3.3. General observations

The 2 1/2 D model of the viscous wall region is probably more useful for depicting the time dependent interaction of near-wall streamwise vortices with riblets than in providing statistics that can be compared against experimental measurements. Figure 5 shows various individual time frames that show some of these interactions. These figures display the cross-stream velocity vectors at various t^+, and the colours represent either downstream momentum or pressure, as indicated.

It is difficult to convey the dynamics of the vortex evolution using static pictures. A brief description is provided of the vortices and how they interact with the riblets; it is hoped that the dynamics of the motion, captured on video, can be appreciated by the reader. As the transient vortices develop close to the wall, intermittent, rapidly decaying secondary flows can easily be observed within the riblet valleys. These are even more apparent in the valleys of U-groove riblets. This is essentially the same phenomenon that Bacher and Smith (1985) suggested as a mechanism in the weakening of the original primary vortices; however, the computationally observed vortices are generated and driven by the periodic outer edge conditions, so the strength of these primary vortices would not be expected to be changed by the riblets. This observation appears to be confirmed by the recent experimental data of Tardu (1991) and Tardu, Truong and Tanguay (1992).

The other significant effect that can be observed is the influence of riblets on the inflows toward the wall that occur beside and between vortices. These inflows carry high streamwise momentum fluid towards the wall and, when reaching a flat wall, spread laterally. This covers an extended area of the surface with a high streamwise velocity gradient and, consequently, a high wall shear stress. The riblets appear to impede this lateral motion of the inrush at the wall. The size of the region with high wall shear stress is then decreased. With the relatively large riblets being used ($h^+ = 16.7$, $s^+ = 33.3$), this area is often confined to a single riblet. The impact of an inflow is most noticeable as an increase in the pressure at the wall (see U-groove riblet pressure contours, last frame of Fig. 5) and is often present only in the riblet immediately below the inflow. Even within riblets directly beneath such inrushes, the riblet valleys retain most of their low momentum fluid. The main increase in wall shear stress appears to occur, for the most part, on the riblet peaks. So, not only is this inflow confined in the spanwise direction, it is also effectively unable to influence large areas of the surface within the riblet valleys.

4. Conclusion and recommendations for future work

A 2 1/2 D model of the viscous wall region has been used to investigate the flow over surfaces with both V-groove and U-groove riblets. The streamwise mean velocity profiles were found to agree with the experimental results of Vukoslavĉević et al.

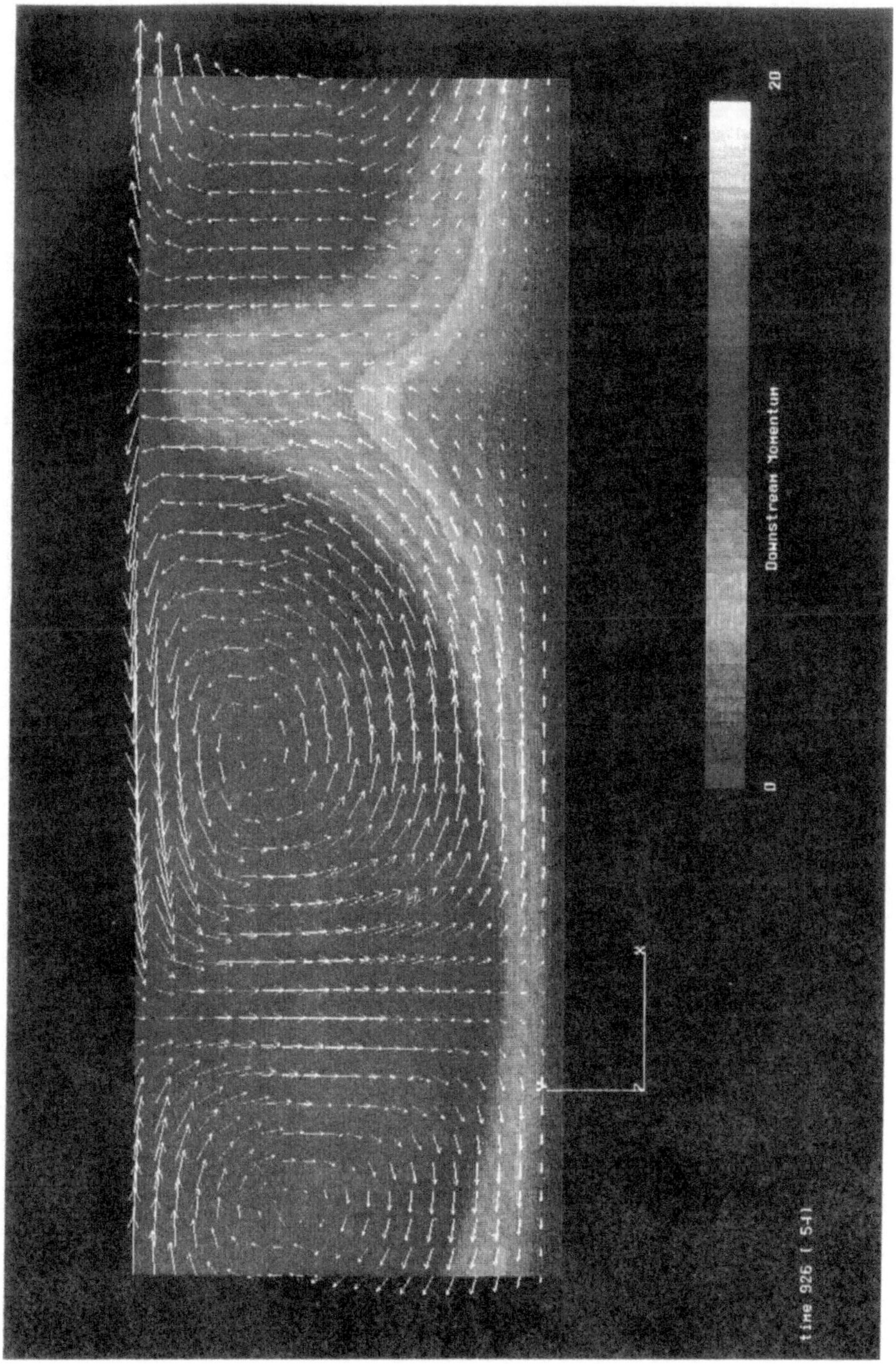

Fig. 5. For legend see p. 311.

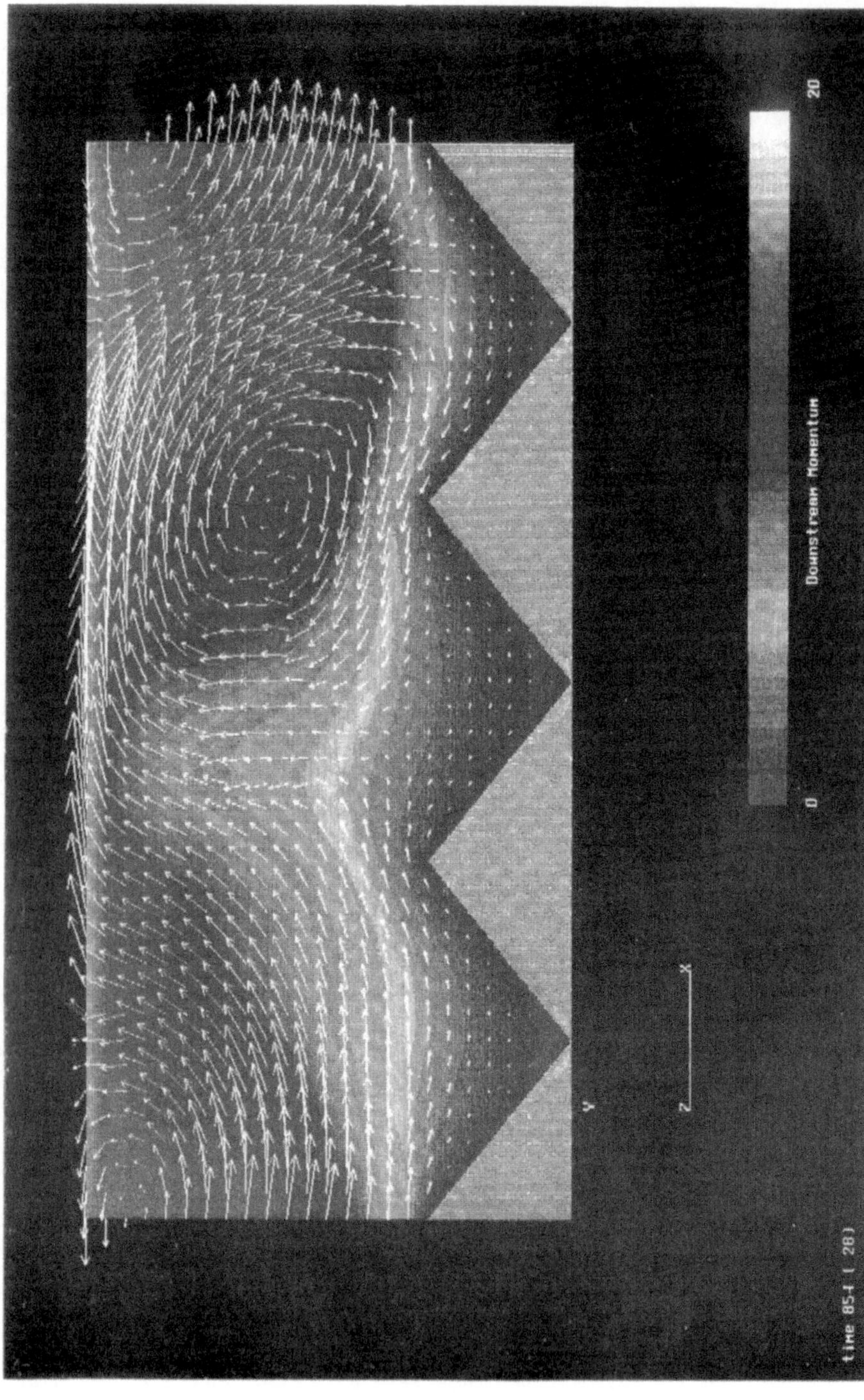
Downstream Momentum
0
20
y
z
x

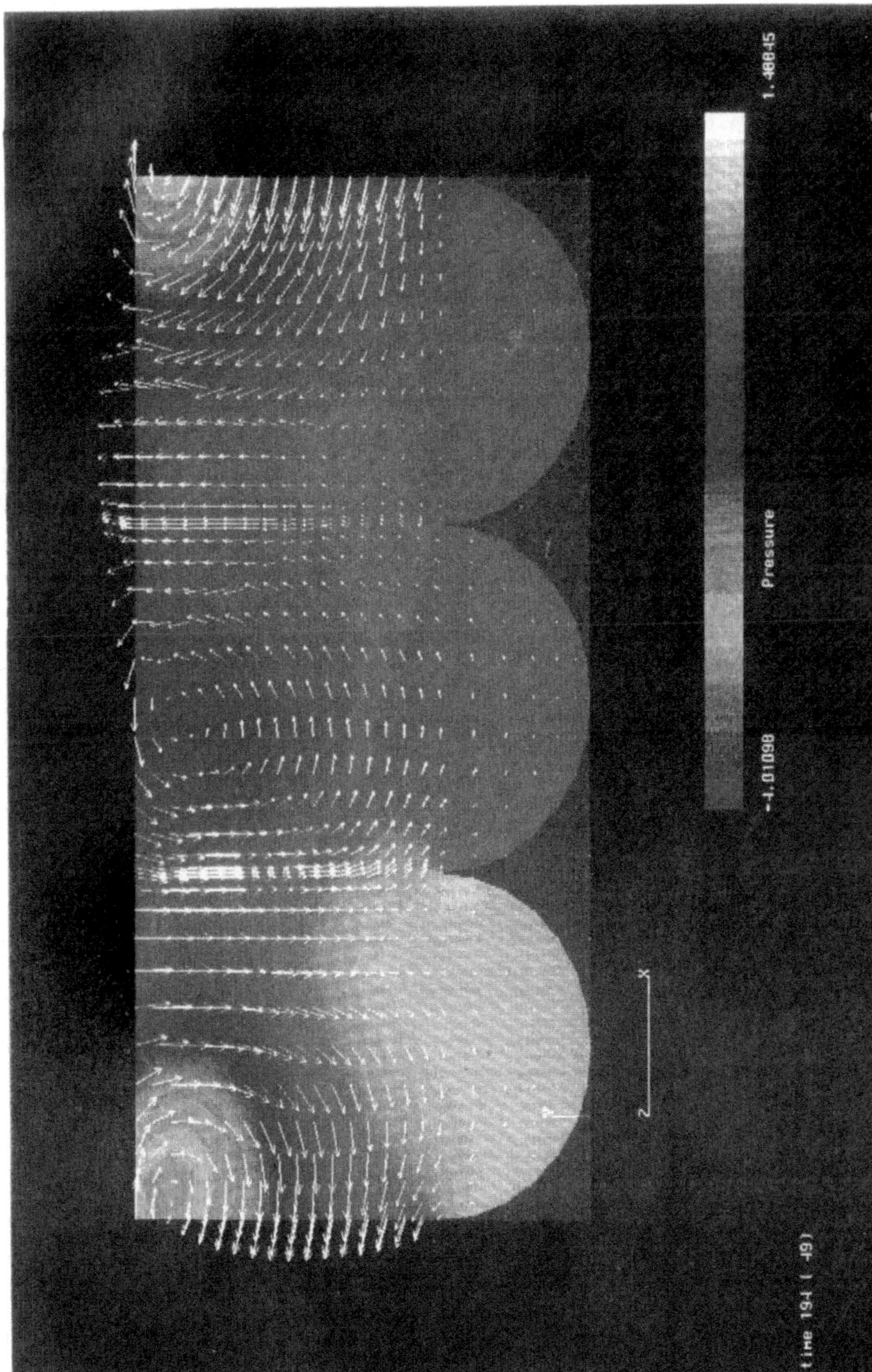

Fig. 5. Calculated flow over plain wall, V-groove and U-groove riblet wall. Vectors are cross-stream vectors; colours represent contours of down-stream or axial momentum or static pressure, as noted. Note the time marks 926, 854 and 194 represent t^+ values.

(1987), showing increases in the mean velocity gradient over the riblet peaks together with large decreases in the riblet valleys. The calculations of the streamwise velocity fluctuation intensities show a comfortable degree of agreement with the limited experimental data available. Visual observations of the transient flows suggest that the riblets interact with the near-wall streamwise vortices weakening them by the generation of transient secondary vortices in the riblet valleys in a manner similar to that proposed by Bacher and Smith (1985). The riblets also appear to limit the lateral spread of inrushes towards the walls and retain low momentum fluid in the riblet valleys effectively isolating much of the wall surface from such inrushes.

The riblets used in this study are not small enough to be drag reducing but were chosen to be approximately the same size as the riblets used in detailed experimental studies. A repetition of this study with smaller, drag-reducing riblets would be beneficial. The increased number of riblets in the domain would also provide better statistics for the various profiles. Comparisons with the very recent direct numerical simulation of the flow in a channel with riblets by Chu (1992) would then be possible. (Note added in proof: these calculations have been completed, see Tullis and Pollard, 1993.)

5. Acknowledgements

The authors would like to extend their sincerest appreciation to Dr. A.M. Savill for the presentation of this work at the 6th European Drag Reduction Meeting, Eindhoven, November, 1991 and for his continued interest and encouragement in this work.

The financial support of the Natural Science and Engineering Research Council of Canada is greatfully acknowledged.

References

Bacher, E.V., Smith, C.R., A combined visualization-anemometry study of the turbulent drag reducing mechanisms of triangular micro-groove surface modifications. AIAA Paper 85-0548 (1985).

Bakewell, H.P., Lumley, J.L., Viscous sublayer and adjacent wall region in turgulent pipe flow. *Physics Fluids* 10 (1967) 1880.

Baliga, B.R., A control-volume based finite element method for convective heat and mass transfer. Ph.D. Thesis, Dept. of Mech. Eng., University of Minnesota (1978).

Baliga, B.R., Patankar, S.V., A control volume based finite element method for two-dimensional fluid flow and heat transfer. *Numerical Heat Transfer* (1983) 245–261.

Bechert, D.W., Bartenwerfer, M., The viscous flow on surfaces with longitudinal ribs. *J. Fluid Mech.* 206 (1989) 105–129.

Benhalilou, M., Anselmet, F., Liandrat, J., Fulachier, L., Experimental and numerical investigation of a turbulent boundary layer over riblets. *Proc. of the 8th Symp. on Turb. Shear Flows,* Munich, Sept. 9–11, 1991.

Chapman, D.R., Kuhn, G.D., Two-component Navier-Stokes computational model of viscous sublayer turbulence. AIAA Paper 81-1024 (1981).
Chapman, D.R., Kuhn, G.D., The limiting behavior of turbulence near a wall. *J. Fluid Mech.* 170 (1986) 265–292.
Choi, K., The wall pressure fluctuations of modified turbulent boundary layer with riblets. *Proc. IUTAM Symp. on Turbulent Management and Relaminarization*, Bangalore. Springer-Verlag (1987) p. 109.
Choi, K., Near-wall structure of a turbulent boundary layer with riblets. *J. Fluid Mech.* 208 (1989) 417–458.
Chu, D.C., A direct numerical simulation of laminar and turbulent flow over streamwise aligned riblets. M.Sc. Thesis, Dept. of Mech. and Aerospace Eng., Princeton University (1992).
Clark, D.G., Boundary layer flow visualization patterns on a riblet surface. QMC-EP-1081 Queen Mary and Westfield College, University of London (1989).
Clark, J.A., A study of turbulent boundary layers in channel flow. *Trans. ASME, J. Basic Engng.* 90 (1968) 455–465.
Djenidi, L., Squire, L.C., Savill, A.M., High resolution conformal mesh computations for V, U or L groove riblets in laminar and turbulent boundary layers. In: Choi (ed.), *Proceedings, 5th European Drag Reduction Working Meeting,* British Maritime Technology, Teddington, Middlesex. Dordrecht: Kluwer Academic Publishers (1991).
Djenidi, L., Riblet flow prediction with a low-Reynolds-number $k-\varepsilon$ model. 6th European Drag Reduction Working Meeting, abstract (1991).
Gallager, J.A., Thomas, A.S.W., Turbulent boundary layer characteristics over streamwise grooves. AIAA paper 84-2185 (1984).
Hatziavramidis, D.T., Hanratty, T.J., The representation of the viscous wall region by a regular eddy pattern. *J. Fluid Mech.* 95 (1979) 655–679.
Kline, S.J., Reynolds, W.C., Schraub, F.A., Runstadler, P.W., The structure of turbulent boundary layers. *J. Fluid Mech.* 30 (1967) 283–325.
Laufer, J., The structure of turbulence in fully developed pipe flow, NACA Report 1174 (1954).
Launder, B.E., Li, S.-P., Prediction of drag reduction by riblets. 6th European Drag Reduction Working Meeting, abstract (1991).
Nikolaides, C., A study of the coherent structures in the viscous wall region of a turbulent flow. Ph.D. Thesis, Dept. of Chem. Eng., University of Illinois, Urbana (1984).
Patankar, S.V., *Numerical Heat Transfer and Fluid Flow.* Hemisphere Publishing Co. (1980).
Prakash, C., Patankar, S.V., A control volume based finite element method for solving the Navier-Stokes equations using equal-order velocity-pressure interpolation. *Numerical Heat Transfer* 8 (1985) 259–280.
Robinson, S.K., Kline, S.J., Spalart, P.R., Quasi-coherent structures in the turbulent boundary layer: Part II. Verification and new information from a numerically simulated flat-plate boundary layer. In: Afgan and Kline (eds), *Zoran Zaric Memorial Seminar on Near-Wall Turbulence.* Hemisphere Publishing (1988).
Saabas, H.J., A control volume based finite element method for three-dimensional, incompressible, viscous fluid flow. Ph.D. Thesis, Dept. of Mech. Eng., McGill University (1991).
Savill, A.M., Drag reduction by passive devices – a review of some recent developments. *Proc. IUTAM Symp. on Structure of Turbulent and Drag Reduction.* Zurich, Springer-Verlag (1989) pp. 429–465.
Smith, C.R., Walker, J.D.A., Haidari, A.H., Taylor, B.K., Hairpin vortices in turbulent boundary layers: the implications for reducing surface drag. *Proc. IUTAM Symp. on Structure of Turbulent and Drag Reduction.* Zurich, Springer-Verlag (1989) pp. 51–58.
Tardu, S., Investigation of the structure of the turbulence in an internal flow manipulated by riblets. Report IMHEF T-91-19, Sept. 1991, EPFL, Lausanne, Switzerland.
Tardu, S., Truong, T.V., Tanguay, B., Bursting and structure of the turbulence in an internal flow manipulated by riblets. *Appl. Sci. Res.* (1993).
Tiederman, W.G., The effect of dilute polymer solutions on the viscous drag and turbulence structure. *Proc. IUTAM Symp. on Structure of Turbulent and Drag Reduction.* Zurich, Springer-Verlag (1989) pp. 187–200.
Townsend, A.A., *The Structure of Turbulent Shear Flow*, 2nd edition. Cambridge University Press (1976).
Tullis, S.W.J., Modelling the time dependent flow over riblets in the near wall region, M.Sc. Thesis, Department of Mechanical Engineering, Queen's University at Kingston (1992).

Tullis, S., Pollard, A., A numerical investigation of the turbulent flow over V and U groove riblets using a viscous wall region model. In: So, Speziale, Launder (eds), *Proceedings, Near Wall Turbulent Flows.* Elsevier (1993) pp. 761–770.

Ueda, H., Hinze, J.O., Fine-structure turbulence in the wall region of a turbulent boundary layer. *J. Fluid Mech.* 67 (1975) 125–143.

Vukoslavĉević, P., Wallace, J.M., Balint, J.-L., On the mechanism of viscous drag reduction using streamwise aligned riblets: a review with new results. *Proc. RAeS Int. Conf. on Turbulent Drag Reduction by Passive Means.* London, RAeS 2 (1987) pp. 290–309.

Walsh, M.J., Riblets, Viscous drag reduction in boundary layers. In: Bushnell, D.M. and Hefner, J. (eds), *Progress in Astronautics and Aeronautics*, Vol. 123 (1990).

Wilkinson, S.P., Anders, J.B., Lazos, B.S. and Bushnell, D.M., Turbulent drag reduction research at NASA Langley – progress and plans. *Proc. RAeS Int. Conf. on Turbulent Drag Reduction by Passive Means.* London, RAeS 1 (1987) p. 1.

Applied Scientific Research **50:** 315–324, 1993.
K. Krishna Prasad (ed.), *Further Developments in Turbulence Management.*

Possibility of drag reduction using *d*-type roughness

K.-S. CHOI[1] & N. FUJISAWA[2]

[1]*Department of Mechanical Engineering, The University of Nottingham, University Park, Nottingham NG7 2RD, U.K.*
[2]*Department of Mechanical Engineering, Gunma University, Kiryu 376, Japan*

Abstract. An experiment was carried out in a low-speed wind tunnel to study the turbulence structure of the boundary layer over a two-dimensional square cavity on a flat plate. The main purpose of this investigation is to examine the way a square cavity modifies the near-wall structure of the turbulent boundary layer leading to a possible drag reduction over *d*-type roughness. The experimental results on pressure coefficient and friction coefficient indicated a small reduction in total drag in this configuration. This seems to be due to the stable vortex flow observed within the cavity which absorbs and reorganizes the incoming turbulence in the cavity, thereby modifying the near-wall turbulence structure of the boundary layer. The resultant turbulence structure was very similar to that over drag-reducing riblets surface.

1. Introduction

In the past decade there has been continuous research efforts to study turbulent drag reduction using passive devices, such as riblets and LEBUs. The majority of the results so far show that up to eight percent of drag reduction can be obtained from riblets, which is believed to be a direct consequence of the modification of near-wall turbulence structure by riblets [1]. At present, riblets are very close to industrial applications for saving fuels and improving aerodynamic performance of aircraft, ships and ground vehicles [2].

Recently, a possibility of drag reduction using *d*-type roughness was discussed by Tani et al. [3] who demonstrated experimentally that the turbulent skin friction over a sparsely spaced two-dimensional square grooves is less than that over a smooth surface for a lower range of the Reynolds number based on the streamwise distance x,

$$R_x < 1.6 \times 10^6 \tag{1}$$

and for the following range of groove spacings,

$$10 < L/d < 40, \tag{2}$$

where L is the groove spacing and d is the depth (see Fig. 1). They have obtained these results during an investigation of the equilibrium turbulent boundary layer over *d*-type roughness [4]. All the equilibrium flows currently known to us have been predicted by Rotta [5] on dimensional ground except for the flow over the *d*-type roughness. It seems natural to expect that the stable vortices inside the grooves must have contributed to the reduction of drag over *d*-type roughness. However, a simple

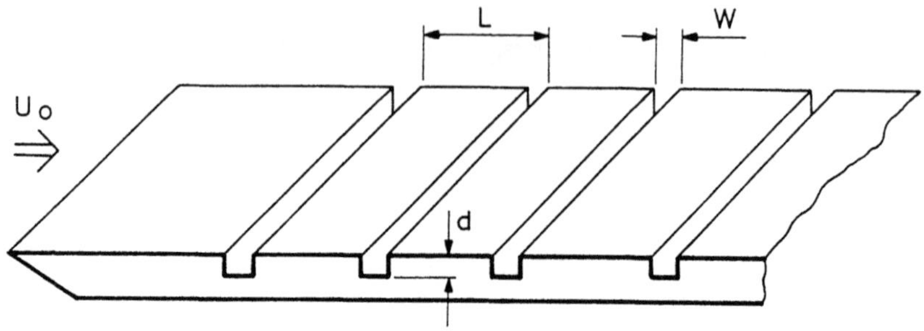

Fig. 1. d-type roughness over a flat plate.

argument that there will be less skin friction because of less surface area does not work here; the experimental results of Tani et al. [3] indicate that a greater reduction in the turbulent skin friction is observed when the spacing between the grooves is increased.

There have been a number of studies in the past on the turbulent boundary layer over a square cavity. Townes and Sabersky [6] have carried out an extensive flow visualization studies in water, where they found that the flow in the cavity was stable only when the Reynolds number based on the cavity depth d and the friction velocity u^* is greater than 150,

$$d \cdot u^*/\nu > 150. \tag{3}$$

Maul and East [7] have also conducted flow visualization with cavities of different aspect ratio. They concluded that the two-dimensional vortex inside the cavity can only exist when the ratio of cavity depth d to the width w is nearly unity,

$$d/w = 1.0. \tag{4}$$

Haugen and Dhanak [8] have pointed out, again from flow visualization studies, that the ratio of the cavity depth d to the boundary layer thickness δ is an important parameter as this can be considered as a measure of the outer-layer influence on the cavity flow. They suggested that this ratio should be much smaller than unity,

$$d/\delta \ll 1, \tag{5}$$

in order for the turbulent drag to be reduced over a cavity.

The main purpose of the present experimental investigation is to study the near-wall turbulence structure over a single two-dimensional square cavity as an effort to identify possible mechanisms for the turbulent drag reduction by d-type roughness observed by Tani et al. [3]. The relaxation process of the turbulent boundary layer after a single cavity is also studied to find out an optimum groove spacing for d-type roughness.

2. Experiments

All the experiments were carried out in an open-return low-speed wind tunnel at the British Maritime Technology, Teddington. The tunnel has a test section 6 m long, 1.22 m wide and 0.6 m high, in which a flat plate is placed at one-third height from the tunnel floor. Attached to the end of the flat plate is a short flap section, which was adjusted by a small angle to avoid any leading-edge flow separation. A trip wire of 2 mm diameter was affixed near the leading edge of the flat plate in order to fix the transition point. Two two-dimensional square cavities were set into the flat plate, one with 10 mm square for the mean flow and turbulence measurements at 1.15 m from the leading edge of the plate, and the other with 25 mm square for the flow visualization studies at 4.66 m. The pressure gradient along the flat plate was zero.

We have measured the static pressures inside the cavity in order to obtain the pressure drag associated with it. There are in total 42 static pressure taps of 0.2 mm diameter in the upstream and downstream faces of the square cavity, and the static pressures were measured using a pressure transducer through a Scanivalve. The accuracy of the pressure measurement was within 0.005 mm Aq. The mean flow and turbulence quantities were measured using DISA 55 M hot-wire anemometer with a single miniature wire. All the data were sampled with PDP 11/44 mini-computer through an A/D converter at a sampling rate of 2 KHz. The tunnel speed was set to 7 m/s throughout the experiments except for the flow visualization studies for which a slower speed of 2 m/s was chosen to obtain good images. Further experimental details and data processing techniques are described elsewhere [1, 9].

In the present experiment, the flow parameters and groove dimensions were chosen to satisfy all the conditions (1) to (5) given in the previous section to make sure that we are in the drag-reduction regime with a stable flow inside the cavity. Table 1 shows our experimental conditions which are compared with those by Tani et al. [3].

Table 1. Experimental conditions and flow parameters.

	Present study	Tani et al. [3]
U_0 [m/s]	7.0	13.0
R_x	0.6×10^6	$0.8\text{–}4.5 \times 10^6$
d^+	210	180
d/δ	0.4	0.24
d/w	1.0	1.0

3. Results and discussions

The distribution of pressure coefficient Cp measured on two vertical faces of the cavity is shown in Figure 2. The x-axis of this figure is the non-dimensional height measured

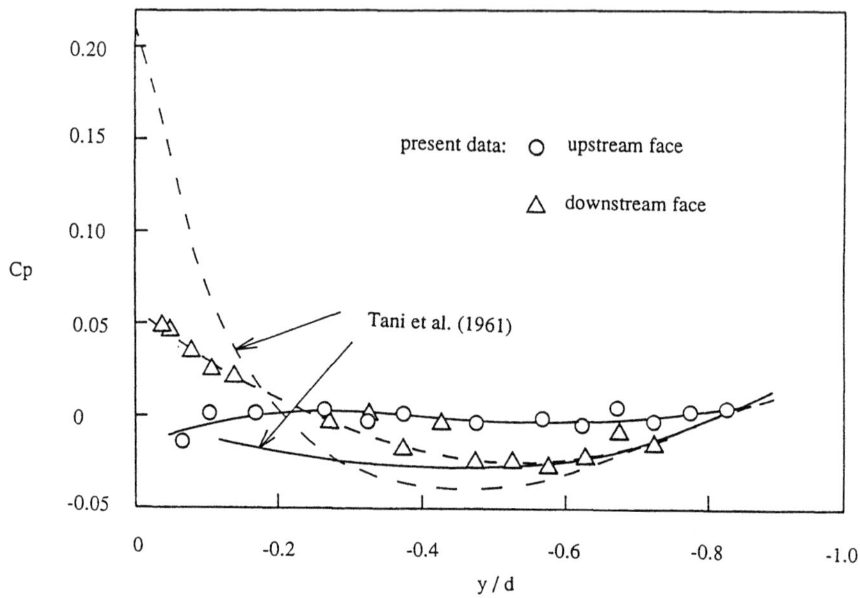

Fig. 2. Distribution of pressure coefficient inside the cavity.

from the top of the cavity, therefore the figure shows the static pressure distribution from the lip of the cavity ($y/d = 0$) to the bottom ($y/d = -1.0$). The experimental results of Tani et al. [10] are also shown in the figure for comparison. The figure shows that Cp has very large positive values near the lip, and negative values around the middle in the downstream face of the cavity. The pressure coefficient is nearly zero in the upstream face of the cavity. All of these results suggest that there is a stable vortex flow inside the cavity, which is confirmed from our flow visualization studies [9]. The drag coefficient C_d estimated from this pressure distribution is 0.001 which is smaller compared with the previous results by Roshko [11], Tani et al. [10] and Haugen and Dhanak [8] who obtained the drag coefficient of between 0.01 and 0.02. The discrepancy may be due to the difference in the relative size of cavity to the boundary layer thickness. In the present experiment the value for d/δ is 0.4, while all the previous experiments were carried out with very thin boundary layers with d/δ between 1.6 to 10. This is in agreement with the suggestion by Haugen and Dhanak [8] that the pressure drag of a cavity should be reduced with a reduction in d/δ. In other words, the vortex in the cavity is stable when d/δ is small since the influence of large-scale turbulence on the cavity flow is less, which leads to a reduction in pressure drag in the cavity.

The mean velocity profiles over a single cavity are shown in Fig. 3 in a semi-logarithmic plot. The streamwise distance x is measured from the downstream face of the groove, and it is non-dimensionalized with the cavity depth d. The points in the figure indicate the experimental data which are compared with a smooth curve representing the velocity profile over a smooth surface. The straight lines show the law

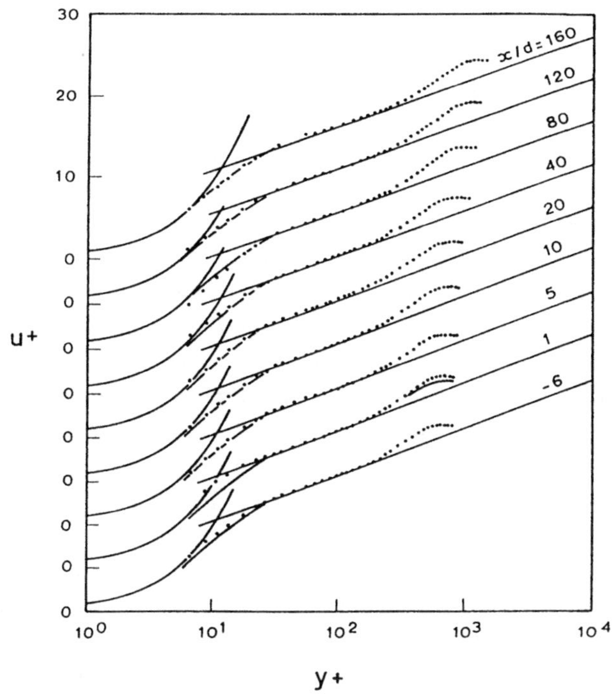

Fig. 3. Logarithmic velocity profiles over a square cavity.

of the wall over a smooth surface suggested by Patel [12]. The deviation from the law of the wall near the outer edge of the boundary layer can be expressed in terms of the law of the wake [13], so that the complete expression for the law of the wall and the law of the wake is given by

$$u^+ = (1/\kappa)\ln y^+ + C + \Pi \cdot W/\kappa, \tag{6}$$

where, κ is the Karman constant ($\kappa = 0.42$) and C is the intercept ($C = 5.45$). The non-dimensional velocity u^+ and the non-dimensional height y^+ are defined by

$$u^+ = U/u^* \tag{7}$$

and

$$y^+ = y \cdot u^*/v, \tag{8}$$

where, u^* is the friction velocity and v is the kinematic viscosity of air. In the law of the wake, Π is the wake parameter and W is the universal wake function. From the bottom of Fig. 3, it is clear that the velocity profile has nearly a decade-wide logarithmic region with a standard buffer and wake region before the cavity at $x/d = -6$. Just after the square cavity at $x/d = 1$, however, this profile changes suddenly in such a way that the velocity near the buffer region and the wake region increases. This behaviour in the mean velocity profile is very similar to the one reported by Fanourakis [14] who tested the *d*-type roughness in a high-speed wind tunnel at the Mach number of 0.78.

The changes in the mean velocity profiles in the near-wall region and the wake region should be reflected in the changes in the friction coefficient C_f and the wake parameter Π, respectively. In order to study the relaxation process of the turbulent boundary layer after a single square cavity, we have obtained C_f from the Clauser plot by fitting the mean velocity to a known logarithmic law for a smooth surface [12]. The wake parameter Π was then found by fitting the entire velocity profile to the expression (6). No attempts were made to account for any differences in the mean velocity profiles due to the presence of a cavity in the boundary layer. In other words, we did not introduce either the roughness shift nor virtual origin in the calculation of these values. This is because there was only a single cavity in the entire flow field so that the boundary layer could not be considered to have a fully-rough velocity profile. Figure 4 shows the change in the skin friction coefficient as a function of non-dimensional distance x/d. The figure also shows the skin friction coefficient for the smooth surface without cavity. The reduction in the skin friction coefficient after the square cavity is evident from this figure, which lasts for more than $x/d=100$, equivalent to $x=40\delta$. The total skin-friction reduction shown here is about one percent, so that we seem to have a small net drag reduction by a single cavity as the pressure drag is negligible. We have to take this result with caution, however, since the error in the calculation of C_f and C_d is estimated to be of the same order of the net drag reduction.

The change in the wake parameter is given in Fig. 5, which also shows an increase after the cavity. Tani showed in his last paper [15] that the skin-friction reduction can

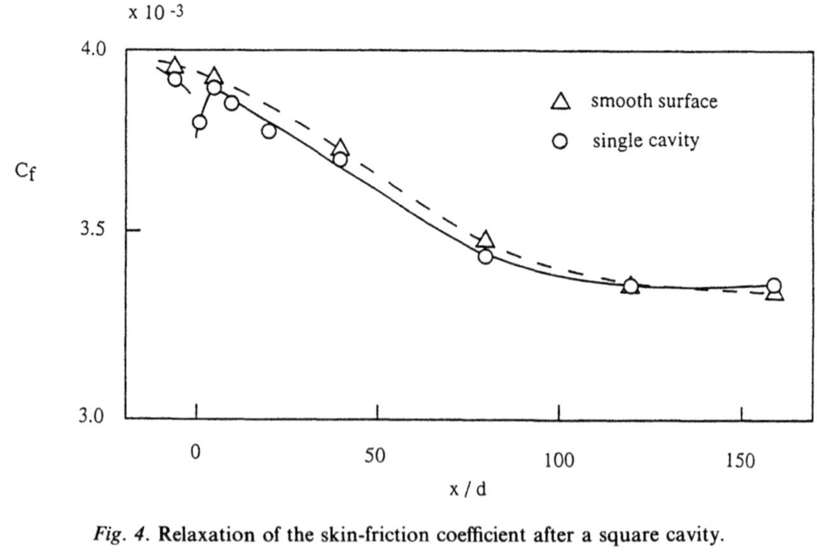

Fig. 4. Relaxation of the skin-friction coefficient after a square cavity.

be achieved by an increase in the wake parameter. He also showed that the turbulent boundary layer over riblets surface has a wake function greater than that over a smooth surface. In this sense, the increase in the wake function observed after the square cavity is consistent with the reduction in the skin friction as shown in Fig. 4. The relaxation length of the wake parameter is around $x/d = 100$, similar to the relaxation length of the skin friction. There was some evidence in the present experimental data that the shape factor is reduced after the square cavity. This is due to the relative reduction of the displacement thickness to the momentum thickness. The shape factor recovers to the smooth-surface value within the distance of $x/d = 100$, similar to the relaxation distance of the skin friction coefficient and the wake parameter. However, an expected reduction in the momentum thickness associated with the reduction in skin friction was not quite evident.

Figure 6 shows the turbulence intensity profiles of the boundary layer on a flat plate with a single square cavity. The points in the figure show the experimental data which are compared with the smooth-surface data indicated by a smooth curve. The present smooth-surface data agrees very well with the Klebanoff's data [16]. It is shown in the figure that the turbulence intensity of the boundary layer immediately above the cavity ($x/d = -0.5$) is larger than the corresponding smooth-surface value. This seems to be due to the effect of turbulent mixing layer developed from the cavity lip in the upstream face. In other words, the near-wall region of the turbulent boundary layer immediately above the cavity is imbedded in the high turbulence region of the mixing layer. After the cavity, the turbulence intensity is reduced as much as 10 percent close to the wall surface, but there was no noticeable changes in the turbulence intensity in the outer part of the boundary layer. This behaviour in the change of turbulence intensity profile over a cavity is very similar to the one over the drag-reducing riblets surface [1].

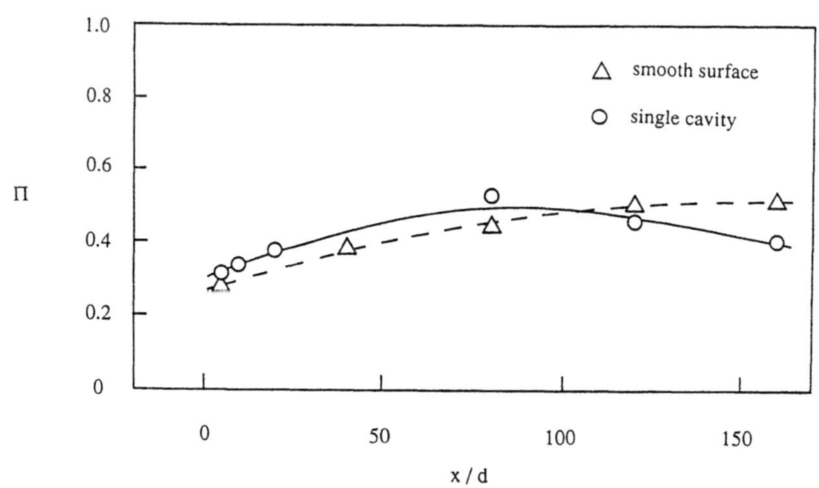

Fig. 5. Relaxation of the wake parameter after a square cavity.

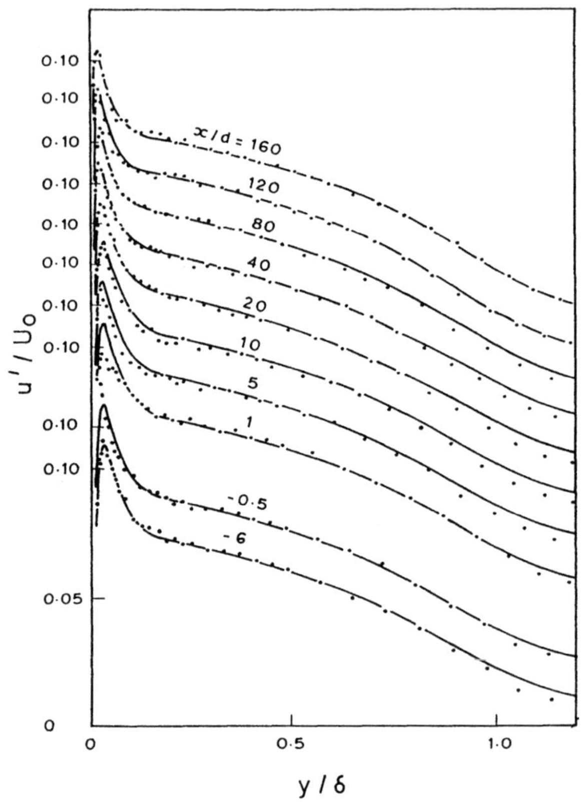

Fig. 6. Turbulence intensity profiles over a square cavity.

Figure 7 shows the energy spectra of streamwise velocity fluctuations measured at several streamwise locations. These spectra were taken at the position where the turbulence intensity is maximum (see Fig. 6). The effects of square cavity on the spectra seem to be to increase the high-frequency component of the energy and to reduce the low-frequence component. The increase in the high-frequency component of the energy seems to be related to the increase in the turbulence intensity (Fig. 6) at $x/d = -0.5$, which is due to the turbulent mixing layer developed from the cavity lip.

4. Concluding remarks

An experimental investigation of the turbulent boundary layer over a single two-dimensional square cavity was carried out by measuring the static pressure distribution inside the cavity as well as the mean velocity and turbulence intensity profiles. The main purpose of this investigation was to examine the way the cavity modifies the near-wall turbulence structure of the boundary layer leading to a possible drag reduction using *d*-type roughness. From this study, it was found that a small amount

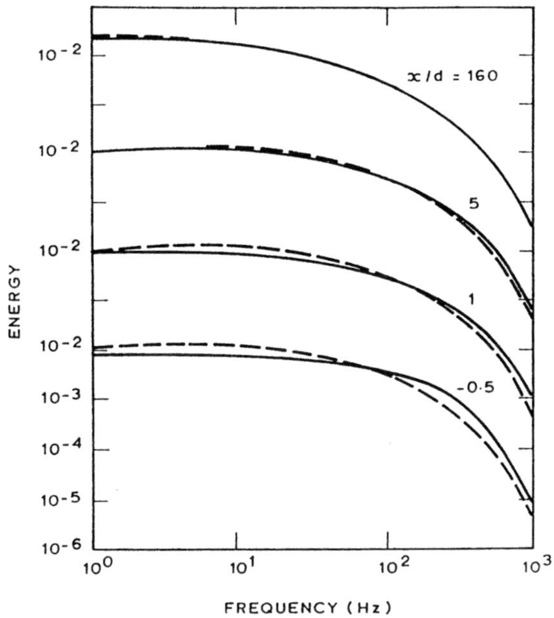

Fig. 7. Energy spectra of streamwise velocity fluctuations over a square cavity.

(of the order of one percent) of net drag reduction can be obtained from a single square cavity. This net drag reduction was mainly due to the reduction in skin friction over the cavity, which lasted for more than $x/d = 100$ (equivalent to $x/\delta = 40$). The pressure drag of the cavity was negligibly small. It was also observed from the present results that the changes in the mean velocity profile, turbulence intensity profile and the energy spectrum over the square cavity were very similar to the corresponding changes over drag-reducing riblets surface [1].

The relaxation study of the flow parameters indicated that the skin-friction coefficient and the wake parameter took more than $100 \cdot d$ to recover after the changes were imposed by the square cavity. This relaxation distance should be considered to be an upper limit of the spacing for the *d*-type roughness in order to obtain the net drag reduction. According to the recent experiments conducted by Tani et al. [17], the optimum groove spacing for the *d*-type roughness is around $20 \cdot d$.

As for the possible mechanisms of turbulent drag reduction using *d*-type roughness, it is believed that they are strongly related to the stable vortex flow inside the cavity, which absorbs and reorganizes the incoming turbulence. The flow visualization studies conducted with a smoke wire and a sheet of laser light [9] indeed confirmed the existence of stable cellular structure within the cavity which has been observed by others [18, 19]. Although the vortex in the cavity rotated slowly at a constant speed most of the time, we could observe accelerations from time to time suggesting that some sort of turbulent events similar to the bursts are taking place near or within the cavity. The frequency of the events was, however, significantly less than that expected from the turbulent boundary layer over a smooth surface. It must be noted that the changes in the near-wall turbulence structure such as the mean velocity and turbulence intensity profiles over the square cavity were remarkably similar to those over the drag-reducing riblets surface [1].

Acknowledgements

This paper is dedicated to the late Professor Tani who provided us with continuous encouragement throughout the course of study.

References

1. Choi, K.-S., Near-wall structure of a turbulent boundary layer with riblets. *J. Fluid Mech.* 208 (1989) 417–458.
2. Choi, K.-S., Cost-effective production of micro-grooves for drag reduction on the surface of aircraft *in situ. SPIE – The International Society for Optical Engineering, Proceedings Volume* 1573 (1992) 136–144.
3. Tani, I., Munakata, H., Matsumoto, A. and Abe, K., Turbulence management by groove roughness. In: Liepman, H.W. and Narasimha, R. (eds), *Turbulence Management and Relaminarisation.* Berlin: Springer-Verlag (1987) pp. 161–172.
4. Tani, I., Some equilibrium turbulent boundary layers. *Fluid Dynamics Research* 1 (1986) 49–58.
5. Rotta, J.C., Turbulent boundary layers in incompressible flow. *Prog. Aeron. Sci.* 2 (1962) 1–219.
6. Townes, H.W. and Sabersky, R.H., Experiments on the flow over a rough surface. *Int. J. Heat Mass Transfer* 9 (1966) 729–738.
7. Maul, D.J. and East, L.F., Three-dimensional flow in cavities. *J. Fluid Mech.* 16 (1963) 620–632.
8. Haugen, R.L. and Dhanak, A.M., Momentum transfer in turbulent separated flow past a rectangular cavity. *ASME J. Appl. Mech.* 33 (1966) 641–646.
9. Choi, K.-S., Fujisawa, N. and Savill, A.M., Studies of drag reducing *d*-type roughness – including flow visualisation, image enhancement and quantitative measurements. In Reznicek, R. (ed.), *Flow Visualisation V.* New York: Hemisphere (1989).
10. Tani, I., Iuchi, M. and Komoda, H., Experimental investigation of flow separation associated with a step or a groove. *Aero. Res. Inst., Univ. Tokyo* 363 (1961).
11. Roshko, A., Some measurements of flow in a rectangular cutout. *NACA TN* 3488 (1955).
12. Patel, V.C., Calibration of the Preston tube and limitations on its use in pressure gradients. *J. Fluid Mech.* 23 (1965) 185–208.
13. Coles, D., The law of the wake in the turbulent boundary layer. *J. Fluid Mech.* 1 (1956) 191–226.
14. Fanourakis, P., A study of rough surfaces with drag reducing properties in turbulent flow. *Engr. Dept. Rept., Cambridge Univ.* October (1989).
15. Abe, K., Matsumoto, A., Munakata, H. and Tani, I., Drag reduction by sand grain roughness. In: Gyr, A. (ed.), *Structure of Turbulence and Drag Reduction.* Berlin: Springer-Verlag (1990) pp. 341–348.
16. Klebanoff, P.S., Characteristics in a boundary layer with zero pressure gradient. *NACA TN* 3178 (1954).
17. Tani, I., private communication (1988).
18. Perry, A.E., Schofield, H.W. and Joubert, P.N., Rough wall turbulent boundary layers. *J. Fluid Mech.* 37 (1969) 383–413.
19. Townsend, A.A., *The Structure of Turbulent Shear Flow*, 2nd edn. Cambridge: Cambridge University Press (1976) pp. 139–143.

Applied Scientific Research **50:** 325–346, 1993.
K. Krishna Prasad (ed.), Further Developments in Turbulence Management.

The drag of three-dimensional rectangular cavities

ERIC SAVORY[1], NORMAN TOY[1], PETER J. DISIMILE[2] & RUSSELL G. DIMICCO[2]

[1]*Fluid Mechanics Research Group, Department of Civil Engineering, University of Surrey, Guildford GU2 5XH, United Kingdom;* [2]*Department of Aerospace Engineering and Engineering Mechanics, University of Cincinnati, Cincinnati, Ohio 45221, USA*

Abstract. Cavities and other surface cut-outs are present on aircraft in numerous forms, from cargo bays and landing gear housing to rivet depressions and panel handles. Although these surface imperfections make a significant contribution to the overall drag on an aircraft, relatively little is known about the flow mechanisms associated with cavities, particularly those which have a strongly three-dimensional geometry. The present work is a wind tunnel investigation of the drag forces and flow regimes associated with cavities having a 2:1 rectangular planform geometry. The effects of both the cavity depth and the flow incidence angle have been examined in terms of the overall cavity drag increment and the mean surface pressure distributions. The drag forces have been determined from both integrated pressures and direct force balance measurements. For the model normal to the flow direction the flow within the cavity was remarkably symmetrical in all the configurations examined. In most cases the cavity flow is dominated by a single large eddy. However, for cavities yawed to other incidence angles there is considerable flow asymmetry, with strong vorticity shedding and high drag in some cases, notably with depth/narrowest width ratio of 0.4–0.5 at 45–60° incidence. The present data correspond well with established results and extend the scope of information available for design purposes and for the development of numerical models.

Nomenclature

A_p = planform area of model
C_D = pressure drag coefficient $(F_D/(A_p \cdot q))$
ΔC_D = drag coefficient increase due to cavity $(C_D - c_f)$
c_f = local skin friction coefficient
C_L = pressure lift coefficient $(F_L/(A_p \cdot q))$
C_p = mean surface pressure coefficient $(P - P_s)/q$
F_D = drag force
F_L = lift force
h = depth
L = longest planform dimension of model
P = surface pressure on model
P_s = freestream static pressure
P_t = freestream total pressure
q = freestream dynamic pressure $(P_t - P_s)$
Re = Reynolds number $(U_R \cdot W/\nu)$
U_R = freestream velocity
W = narrowest planform dimension of model
Z = vertical cartesian coordinate
Ψ = incidence angle
ν = kinematic viscosity

1. Introduction

The reduction of drag forces is one of the major continuing aims of the aeronautics industry. A significant source of parasitic drag is that due to surface excrescences and cut-outs and yet very little is known about the flows associated with such geometries. Indeed, these surface imperfections, for example; roughness, equipment housings, external instruments, panel handles, ridges and holes account for up to 3% of the total drag on a modern aircraft, such as the A320. Hence, any improvements to the external surface profile can lead to significant reductions in fuel consumption and enhancement of operational performance.

Our knowledge of the effects of surface cut-outs and cavities is mainly limited to the overall drag increments due to single or multiple holes, such as the early data obtained by Wieghardt (1946), Tillmann (1951), Roshko (1955) and Tani et al. (1961) using essentially two-dimensional rectangular slots normal to the flow direction. The case of a fully three-dimensional rectangular cavity has been examined by a few workers, notably Friesing (1936), Maull and East (1963), Rossiter (1964) and Gaudet and Winter (1973). Much of the early research was summarized by Hoerner (1965), whilst AGARDograph 264, Young and Paterson (1981), provides an excellent presentation of the best drag data in a form that may be used for design purposes. However, apart from a few surface pressure distribution measurements by a limited number of workers, such as Roshko (1955), Maull and East (1963), Rossiter (1964) and, more recently, Sinha et al. (1982), Kaufman et al. (1983) and Plentovich (1990), the drag data has been derived from force balance measurements which provide very little insight into the flow regimes occurring within the cavities.

The case of the two-dimensional square or rectangular cavity has attracted the attention of several workers developing numerical models because it is one of the simplest boundary value problems that can be posed for the Navier Stokes (NS) equations, Mills (1965). Some workers have developed analytical models, such as Haugen and Dhanak (1966) who treated the turbulent separating shear layer across the top of a rectangular cavity as a two-dimensional plane jet and applied the boundary layer momentum and continuity equations. The results, at Reynolds numbers up to 1.3×10^5, for surface pressures, mean velocities and turbulence intensities showed reasonably good agreement with experimental data. Their findings confirmed that the overall cavity drag decreased with increasing thickness of the approach flow boundary layer. Steady NS solutions have been presented, showing the presence of one primary vortex in the cavity and up to two secondary corner vortices, for Reynolds numbers up to 4×10^2, Burggraf (1966), up to 2.7×10^3, Pan and Acrivos (1967) and up to 5×10^4, Nallasamy and Prasad (1977). Some unsteady NS computations were undertaken by Donovan (1970) at a Re of 10^2 for a square cavity and these showed good agreement with mean velocity data from a steady solution. Solutions for supersonic, high Reynolds number, flows include those by Hankey and Shang (1980), who used NS calculations to determine the pressure oscillations over a cavity for Mach 1.5 and Re of 2.6×10^7, and by Brandeis (1982) who used a compressible boundary layer model above the cavity and the incompressible NS equations within the hole for Mach 2.25 and Re of 10^5. An investigation of the flow in a rectangular cavity at Mach 1.5 and 2.5 was carried out by Zhang and Edwards (1988) using the two-dimensional unsteady NS equations, in terms of mass averaged variables, coupled with a two-layer eddy viscosity turbulence model. Their computations demonstrated the presence of self-sustained flow oscillations and the formation and shedding of leading edge vortices. These oscillations were the result of a feedback interaction in which the shear layer plays the dominant role by driving the leading edge vortices downstream to mix with the trailing edge vortex. The trailing edge vortex movement sends a pressure pulse upstream which deflects the oncoming boundary layer, subsequently forming another leading edge vortex and, thereby, closing the loop. Baysal and Stallings (1988) presented surface pressure distribution results from a two-dimensional NS solution with an algebraic turbulence model up to Mach 1.5 and a *Re* of 6.6×10^6 per m. Whilst the data showed some agreement with experimental measurements, there were significant discrepancies caused by the fact that the experimental cavity was narrow enough to exhibit strong three-dimensional flow effects. Indeed, such three-dimensionality is likely to occur in all but the very widest cavities. Furthermore, even in nominally two-dimensional rectangular cavity flows Maull and East (1963) observed the presence of regular three-dimensional cellular structures which were reflected in the periodic changes in the mean pressure distributions across the base in the lateral direction. Rizetta (1988) carried out some unsteady NS calculations of the flow associated with a three-dimensional cavity under supersonic conditions. The results from two and three-dimensional predictions for a cavity with $W/L = 0.53$ and $h/W = 0.37$ at 90° incidence were presented and compared with experimental data from Kaufman (1983), obtained under similar test conditions of Mach 1.5 and *Re* of 1.09×10^6. The computations showed good agreement for the mean static pressures, whilst the amplitudes of the harmonic frequencies of the cavity acoustic spectra were overpredicted.

The broad categorization of two-dimensional rectangular cavity flows has been illustrated by Plentovich (1990). For deep cavities, which have a width in the flow direction to depth ratio (W/h) less than about 10, the flow is termed 'open' and is dominated by a primary vortex driven by the shear layer passing over the top of the cavity. In the case of shallow cavities the separated shear layer attaches on the cavity base causing a backward-facing step flow downstream of the upstream wall and a forward-facing step flow immediately upstream of the downstream face. This type of flow is termed 'closed' and is characteristic of cavities with streamwise dimension (width) to depth ratios above about 13. However, it should be noted that for three-dimensional cavities closed systems of recirculating flow cannot exist, Hunt et al. (1978), and so the flow regimes are considerably more complicated with fluid entering the cavity and vorticity being shed from it. In the present work all of the cavity configurations were highly three-dimensional with low planform aspect ratios and most of the flows were in the 'open' regime. Detailed surface pressure measurements and overall drag force balance tests were carried out for a range of cavity depth to width ratios using cavities with a 2:1 rectangular planform. One major aspect of the work was the examination of the effects of approach flow incidence direction which had not been examined by previous workers.

2. Experimental details

In this section details are provided concerning the apparatus arrangements for the cavity experiments, together with an outline of the scope of the tests. The wind tunnels and boundary layers are dealt with first and this is followed by descriptions of the models and measurement systems.

2.1. Wind tunnels and boundary layers

Two wind tunnels were used in the experiments. The first tunnel, in the Department of Civil Engineering at the University of Surrey, which has working section dimensions of 1.37 m height by 1.07 m width by 9.0 m length, was employed for all the drag balance measurements. A 20 mm high metal wall fence was placed across the tunnel near the contraction to produce a thick boundary layer. Across the tunnel the velocity was uniform to within $\pm 2\%$, whilst the rms turbulence intensity was within 5%. These boundary layer profiles were obtained at a freestream velocity of 12 m/s which was close to the speed used in most of the subsequent experiments.

The second wind tunnel, in the Department of Aerospace at the University of Cincinnati, has working section dimensions of 0.61 m height by 0.91 m width by 2.44 m length and this was used solely for the rectangular cavity pressure distribution measurements. Here, the model was mounted in the tunnel roof and a much thinner turbulent boundary layer was generated which provided a valuable comparison with the data from the first tunnel experiments. The boundary layer profiles were obtained with a freestream velocity of 7 m/s which was that utilized in the cavity experiments.

The boundary layer parameters for both tunnels are given below.

Drag balance experiments	*Pressure measurements*
$\delta = 320$ mm	$\delta = 26$ mm
$\delta_* = 27.9$ mm	$\delta_* = 4.1$ mm
$\Theta = 21.2$ mm	$\Theta = 2.7$ mm
$c_f = 0.00265$	$c_f = 0.00520$

In each case the boundary layer profiles were measured using a hot-wire anemometer and the skin friction coefficient computed from the standard smooth wall logarithmic law plot.

2.2. Main wind tunnel model

The main rectangular cavity model, constructed for the surface pressure measurement experiments, is of Perspex construction and has a 2:1 planform ratio of dimensions 106 mm (L) by 53 mm (W). There are 0.79 mm diameter pressure tappings over one half of each of the narrow walls, facing each other, and a similar arrangement on the wider walls. On one half of the long walls there are thirty-three rows of tappings each row spaced 5 mm apart beginning at 2.5 mm from the rim. Each row contains six tappings. On the opposite half of each long wall there are two rows of six tappings placed 5 mm apart, commencing at 5 mm from the lip. On each half of the narrow wall there are thirty-three rows of tappings positioned 5 mm apart, starting at 2.5 mm from the rim, each with three tappings. The opposite half of the wall has two rows of three tappings spaced 5 mm apart, beginning at 5 mm from the lip. Within every row on each wall the spacing between each tapping is generally 10 mm although they are more closely spaced near the corners on the long walls. In addition, there are forty-two tappings on one half of the piston-style base, mostly spaced 10 mm apart but with additional taps in the corners.

The cavity is built into a turntable which may be inserted into the tunnel wall to permit precise rotation of the model to the desired incidence angle. In view of the generally small magnitudes of the pressures being measured it was essential to prevent any leakage of air through the sides of the base. This was achieved by taping plumbers white around the base before inserting it into the model cylinder. In addition, all unused pressure tappings were sealed off to prevent leakage through the holes. The model utilized in the drag balance experiments is described later.

2.3. Pressure measurement apparatus

The apparatus arrangement for the mean surface pressure measurements is illustrated schematically in Fig. 1. The model was mounted in the roof, flush with the tunnel interior, at 1.2 m downstream of the contraction and a pitot-static tube was positioned in the freestream, directly across the tunnel at the same downstream location as the cavity, for measurement of the reference dynamic pressure. The large number of pressure tappings on the model were attached to connectors which, in turn, were inserted into Scanivalve switch mechanisms. This allowed the connection of each tapping to one side of a differential pressure transducer. The other, reference, side of the transducer was connected to the freestream static pressure, via a damper to ensure that unwanted fluctuations were removed. A ± 13.6 mm WG range transducer was used in the experiments and this was found to be capable of resolving the large variations in the magnitudes of the pressures, from the high positive values at the reattachment point of the separated shear layer to the very small pressures deeper into the cavity. The freestream dynamic pressure was constantly monitored during the measurements using a second transducer which had a ± 38 mm WG range. A PC was utilized to acquire the pressure data from the two transducers and these were used to compute the pressure coefficients defined as

$$C_p = \frac{P - P_s}{P_t - P_s},$$

where P is the model surface pressure and P_t and P_s are the freestream total and static pressures, respectively. The computer also controlled the operation of the Scanivalve system via a digital control unit. Typically, a delay time of 15 s was used after each switching of a Scanivalve port to allow settling of the pressure at the transducer.

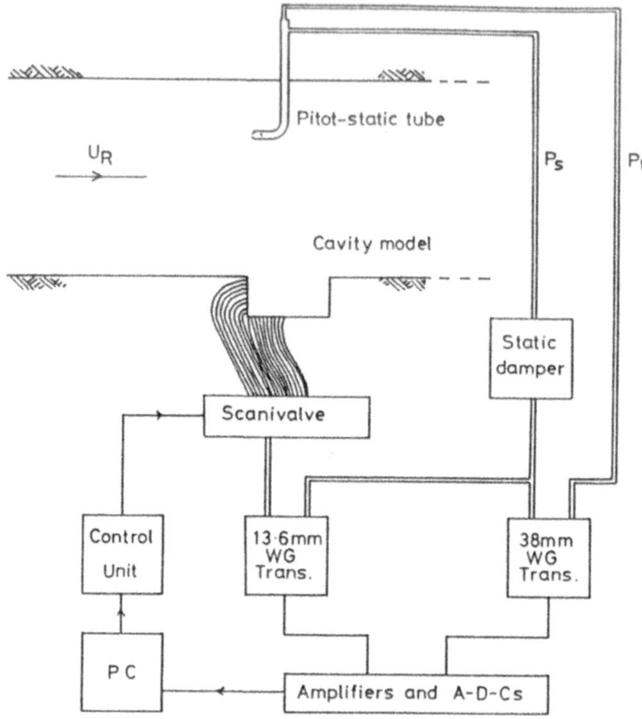

Fig. 1. Schematic arrangement of pressure measurement apparatus.

Similarly, the sampling times were 30 s at each port to permit determination of reliable mean values. The reference dynamic pressure was monitored throughout each experiment and the accuracy of the data was estimated to be ± 0.003 in C_p. In these experiments very small cavities, in the form of pressure tappings, were used to obtain data for a large cavity model. However, an assessment of the errors involved in pressure tap measurements, according to Shaw (1960), revealed that such errors were more than an order of magnitude less than the overall accuracy stated above. The overall lift and drag coefficients, based on the model planform area and the freestream dynamic pressure, were obtained by integrating the surface pressures and resolving the force components in the appropriate direction. For the drag force the integration area was the complete cavity wall, whilst the lift force was computed from the pressure distribution on the cavity base. However, in both cases the cavity base area was used as the normalizing area parameter. The drag increment due to the cavity, ΔC_D, was determined by subtracting the approaching boundary layer skin friction coefficient from the drag coefficient. Complete surface pressure distributions were obtained for the rectangular cavities with h/W from 0.1 to 0.7, in 0.1 intervals, and for incidence angles of 0, 30, 45, 60 and 90°. The freestream velocity was 7 m/s giving a Reynolds number of 2.44×10^4.

2.4. Drag force balance apparatus

In order to obtain direct measurements of the drag coefficients, for comparison with the data from integrated pressure distributions, a force balance was designed and installed into the floor of the 1.37 m by 1.07 m by 9 m tunnel, as illustrated diagrammatically in Fig. 2. The balance has a plan area of 0.5 m length by 0.3 m width into which ground boards containing models can be inserted. Essentially, the balance mechanism is a parallelogram in which rotation can occur at each corner on frictionless bearings. This mechanism is attached to a rigid frame which, in turn, is fastened to the underside of the tunnel. A load cell is incorporated into the framework and a small threaded shaft on one upstream vertical arm of the balance impinges upon the load cell sensor. On the other upstream vertical member is a cantilever upon which weights may be placed to pre-load the load cell. The gap between the balance plate and its surrounding frame at the tunnel floor is of the order of 0.05 mm but the actual position of the plate within the frame is set by adjusting the cantilever loading and the threaded shaft. In order to prevent contact bounce between the load cell and the shaft under dynamic loading a simple damping device is placed in parallel with these two components.

The frame surrounding the balance plate is pressure tapped, with six tappings on

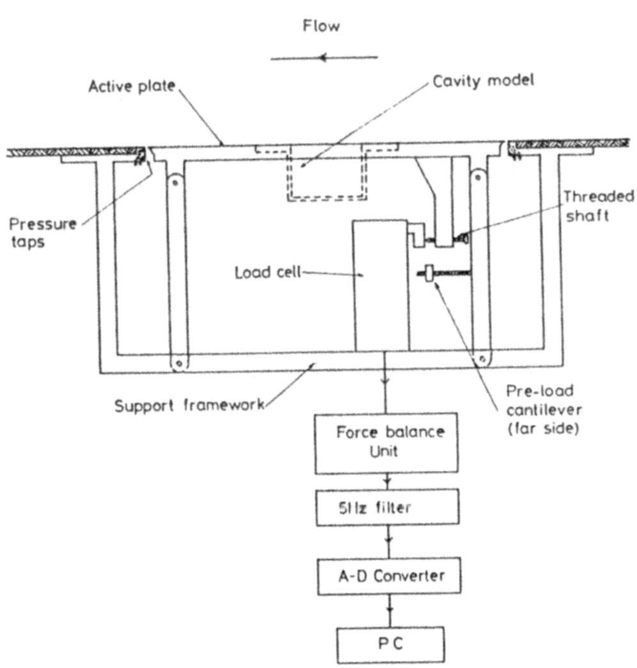

Fig. 2. Diagrammatic layout of drag balance.

both the upstream and downstream faces and two tappings on the two side faces. This was carried out in order to assess the effect of any flow through the gaps on the loading on the sides of the balance plate. However, for the cavity models at least, there is no measurable difference between the pressures on the upstream face and those on the downstream face. The ground board is constructed from timber with a circular hole cut in the centre to allow models to be slotted in and then rotated to the required incidence angle. The board is screwed down onto the plate with packing material underneath to ensure a flush finished surface. The balance is located approximately 6 m downstream of the tunnel contraction. The output from the load cell is taken to a unit which contains a precision high gain amplifier. This provides an accurate output with a calibration of 76.19 g/V and repeated calibration tests using small weights showed that hysteresis was less than ± 0.0002 V or ± 15 mg. Since only mean loads were of interest in the present work the amplified output was filtered at 5 Hz before being sampled by a PC via and A-D converter system. Before each set of experiments the load cell and digitization system were calibrated *in situ* to an accuracy of better than 0.001 V, i.e. 76 mg.

For each model configuration the load with no airflow was measured and this was then subtracted from the load measured at the test velocity. A similar procedure was adopted for measuring the load on the plate under the same freestream conditions but without a cavity present. Hence, unlike the pressure integration approach, this direct method also incorporated the wake drag associated with the cavity. From this data the net drag coefficient due to the cavity, ΔC_D, could be computed by taking the load increment and dividing by the cavity model planform area and the freestream dynamic pressure. A sampling time of 40 s was used for each measurement and several sets of samples were taken for each configuration to ensure reliability of the data and to assess the experimental scatter.

The model used in these experiments was constructed from timber and had a plunger-type base to allow variation of the cavity depth. The planform dimensions were 70 mm (W) by 140 mm (L) and the tests were conducted with a freestream velocity of 11.1 m/s giving a Reynolds number of 5.12×10^4. The experiments were conducted with models of varying depths, from $h/W = 0.05$ to 1.6, at an incidence angle of 0° and a limited number of tests were carried out to assess the effects of changing the incidence angle.

3. Results and discussion

In this discussion the variation in the pressure distributions with depth for the 0° incidence case is examined first. All of the distributions are illustrated such that in each diagram the cavity has been 'cut' vertically at the corners and the wall distributions 'unfolded'. For the shallowest cavity with $h/W = 0.1$ (Fig. 3) the flow regime appears to be closed. The maximum coefficient measured on the base of 0.25 is about the upper limit suggested by Roshko (1955) for rectangular cavity flows. It may

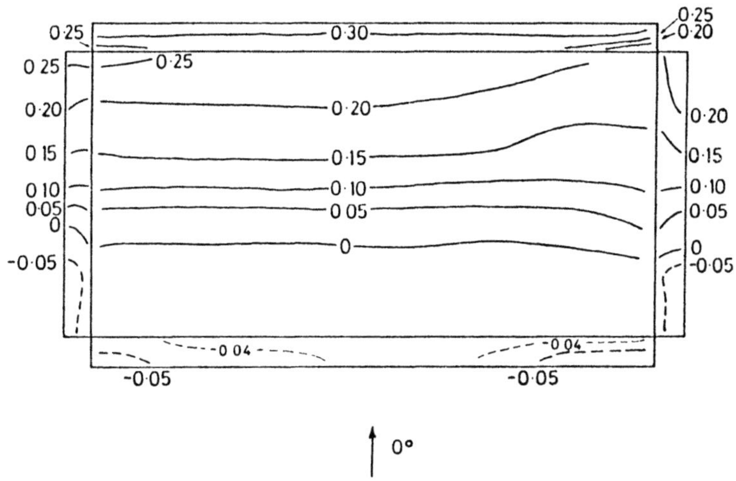

Fig. 3. Pressure distribution with $h/W = 0.1$ for 0° incidence angle (showing closed regime).

be seen that at this incidence angle the flow is remarkably uniform in the lateral direction with no noticeable end effects. Indeed, this spanwise uniformity is in contrast to the series of cell-like structures noted by Maull and East (1963), the number of which was related to an integer multiple of the planform aspect ratio. Their results showed lateral variations in the base pressure distributions associated with these cells which, at a given aspect ratio (L/W), had a fixed, periodic nature. It was found that the cells had a preferred span which was a function of the chord and depth of the cavity. The lack of such structures observed in the present work may be due to the particular planform aspect ratio since Maull and East found that the cells did not occur at every aspect ratio.

With increasing cavity depth the formation of the main vortex becomes more evident such that by $h/W = 0.3$ (Fig. 4) the flow regime is open and the vortex centre is located at $0.74W$ from the leading edge and about $0.5h$ from the base. As the depth becomes greater the vortex moves upstream with the centre located at $0.69W$ from the leading edge at $h/W = 0.4$ (Fig. 5), $0.65W$ at $h/W = 0.5$ (Fig. 6), $0.63W$ at $h/W = 0.6$ (Fig. 7) and $0.61W$ at $h/W = 0.7$ (Fig. 8). At the same time the vortex centre appears to move upwards away from the base of the cavity. As a consequence of these movements the vortex location asymptotes to the position typically associated with a two-dimensional rectangular cavity, i.e. $0.6W$ in the downstream direction and $0.75h$ vertically from the base.

Considering now the case of the cavity at 90° incidence, the shallowest hole with $h/W = 0.1$ (Fig. 9) displays the features of a closed flow regime. Here, the separated shear layer from the upstream wall reattaches on the base at approximately $7h$ from the wall. This value has been estimated by assuming that reattachment occurs a short distance before the peak surface pressure is reached. The flow is, therefore, similar to a backward-facing step regime and the pressure coefficient on the backward face, at

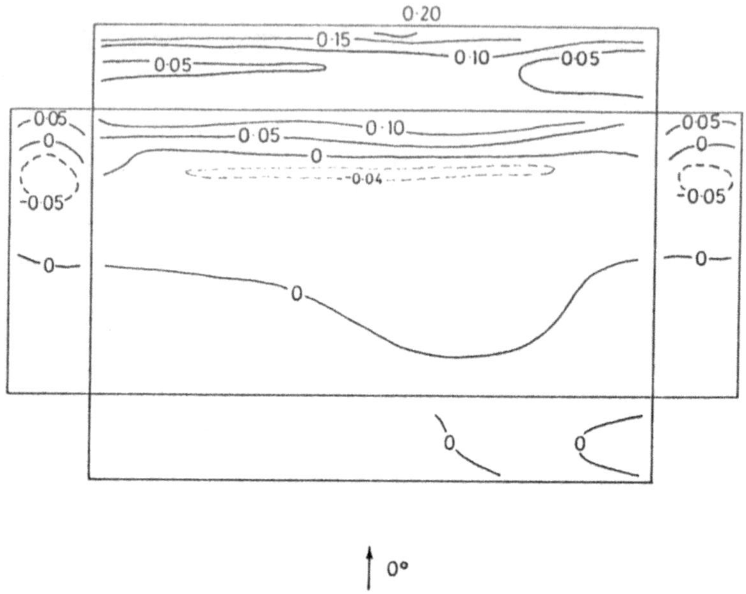

Fig. 4. Pressure distribution with $h/W = 0.3$ for 0° incidence angle (showing open regime).

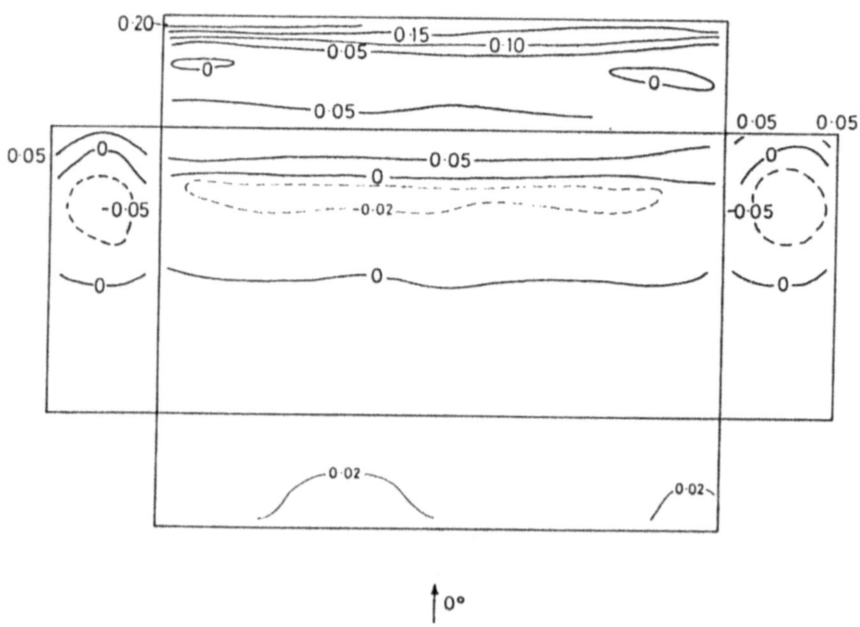

Fig. 5. Pressure distribution with $h/W = 0.4$ for 0° incidence angle.

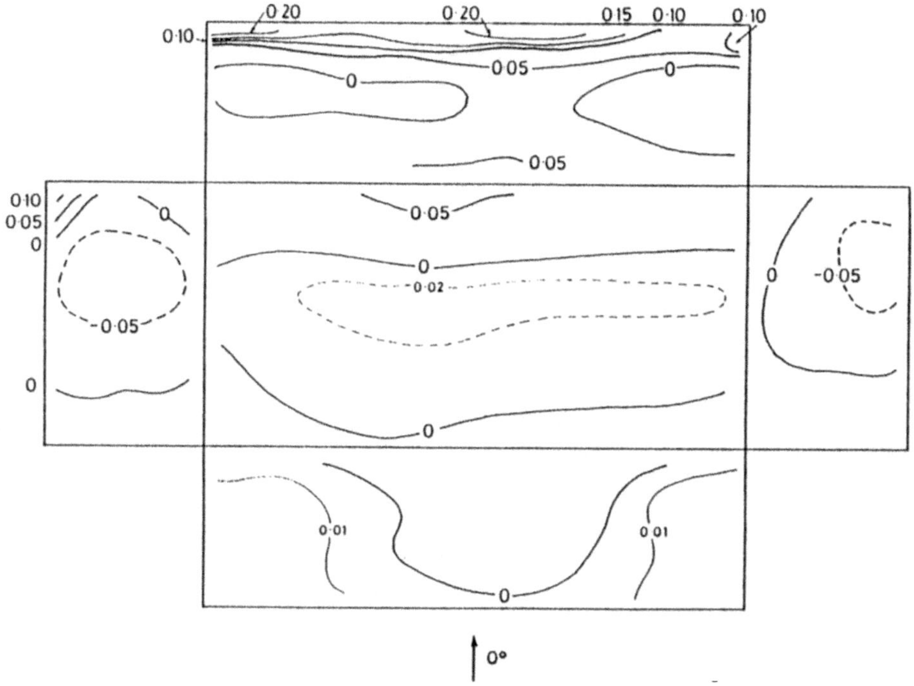

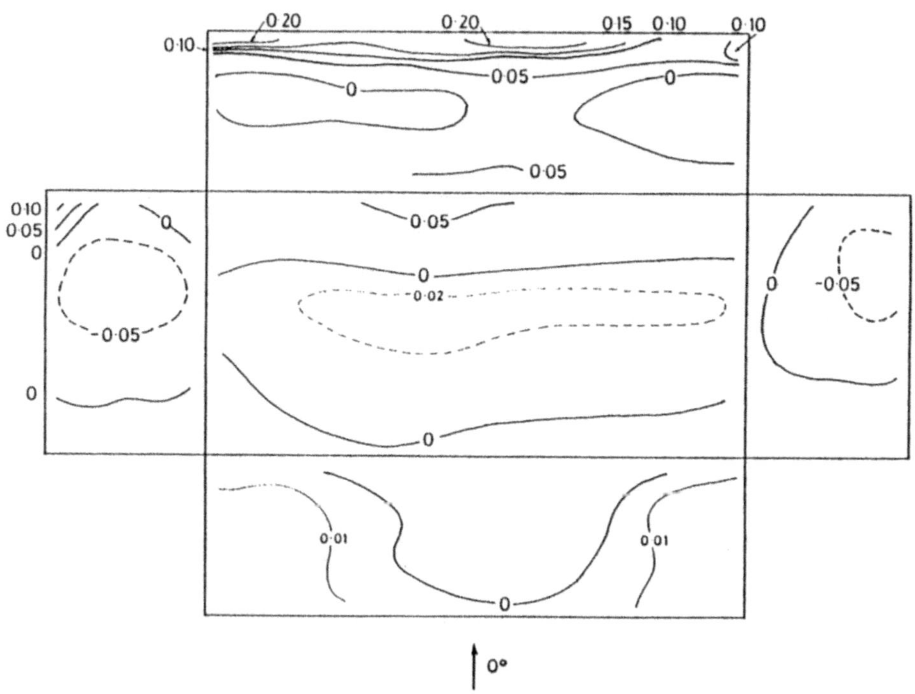

Fig. 7. Pressure distribution with $h/W = 0.6$ for 0° incidence angle.

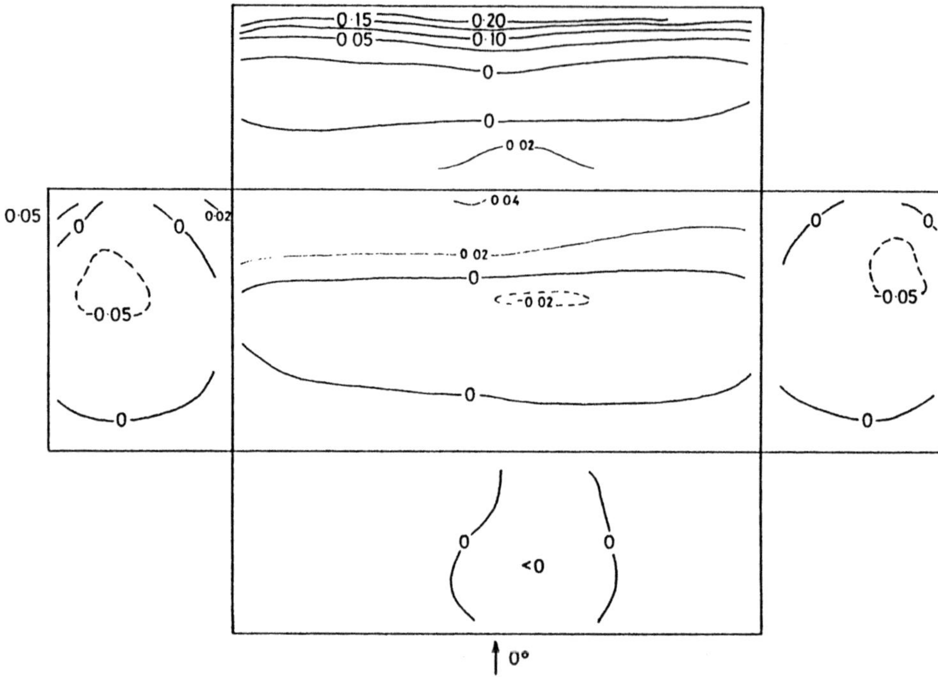

Fig. 8. Pressure distribution with $h/W = 0.7$ for 0° incidence angle.

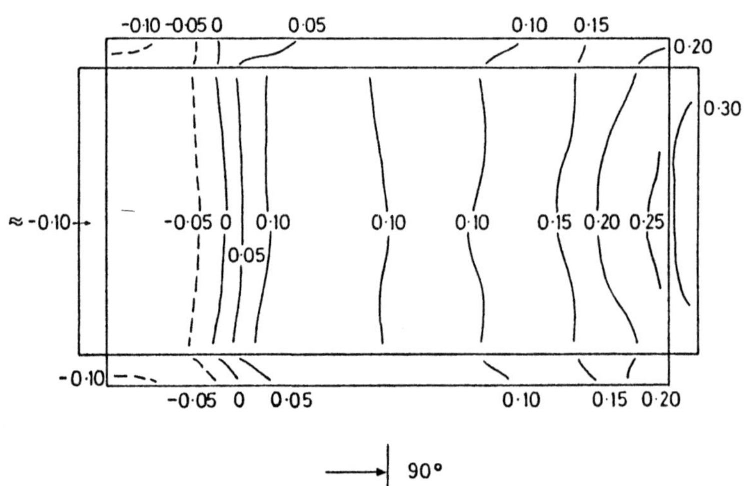

Fig. 9. Pressure distribution with $h/W = 0.1$ for 90° incidence angle (showing closed regime).

−0.10, is also typical of such a flow. Towards the downstream wall the pressure rises rapidly and the flow separates in a similar fashion to that immediately upstream of a forward-facing step.

With a small increase in depth to $h/W = 0.2$ (Fig. 10) the regime has completely changed to the open case with no evidence of early reattachment on the base. As the depth increases, the formation of a negative pressure region across the base and down the wide walls becomes more pronounced and moves downstream. If, as suspected, this region marks the presence of the main cavity recirculation then it would appear that this flow structure originates near the upstream wall in the shallower cavities but is located further downstream as the cavity deepens.

In the deepest cavity with $h/W = 0.7$ (Fig. 11) the separated shear layer reattaches near the lip of the downstream wall and a large symmetric vortex forms beneath it. For this, essentially narrower, cavity the vortex centre location is further downstream, relative to the streamwise dimension, than that noted in the 0° incidence case. It is possible that for greater depths the vortex would move further upstream towards the equivalent two-dimensional cavity location.

Having examined the effects of varying the cavity depth for two symmetric incidence angles, the influence of changing incidence angle for a given depth, namely $h/W = 0.5$, will now be described. The distribution at 0° incidence (Fig. 6) is essentially symmetrical with a well-defined vortex structure, as described earlier. The shape of the distribution along the streamwise centre-line is similar to that measured by Haugen and Dhanak (1966) for a two-dimensional cavity with the same h/W ratio except that the magnitudes of the pressures are slightly lower throughout in the present work. One of the reasons for this discrepancy is that the δ/W ratio in the present work is

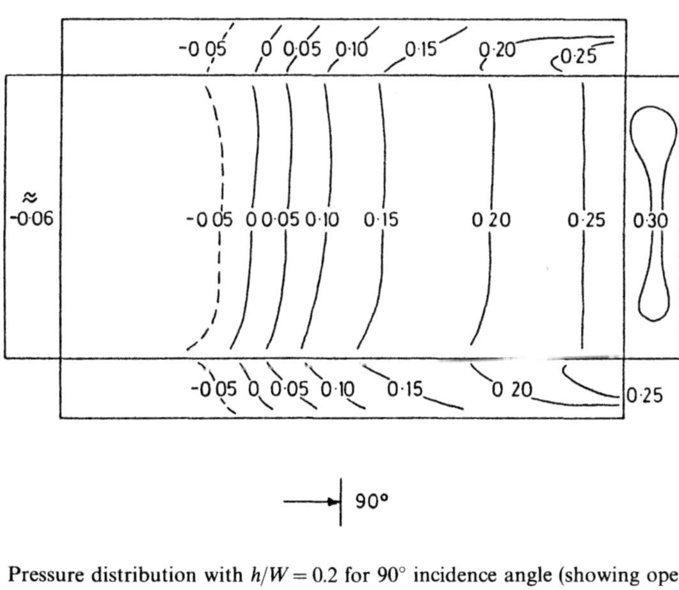

Fig. 10. Pressure distribution with $h/W = 0.2$ for 90° incidence angle (showing open regime).

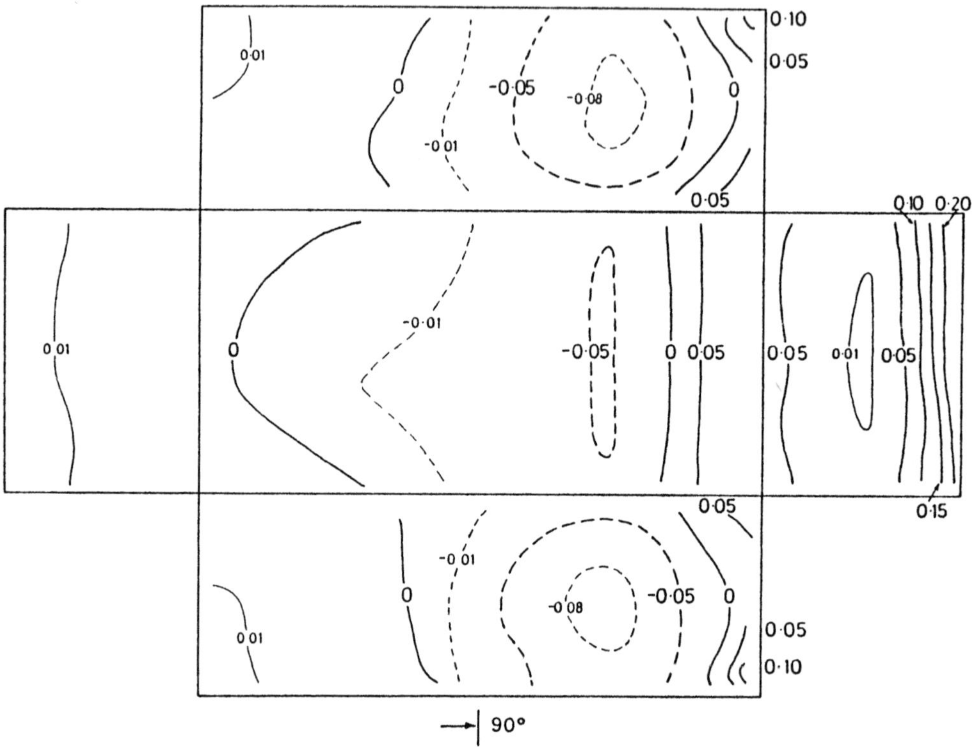

Fig. 11. Pressure distribution with $h/W = 0.7$ for 90° incidence angle.

greater than that used by Haugen and Dhanak. At 30° incidence (Fig. 12) the distribution becomes very asymmetrical although the shear layer still creates a stagnation region on the downstream long wall. The line of the vortex remains approximately parallel to the major axis but the downstream end is raised with a possible consequential increase in vorticity shedding.

At 45° incidence (Fig. 13) the stagnation region is near the top of the cavity at the downstream corner. It would appear that the main axis of recirculation within the cavity is still approximately parallel to the major model axis with strong activity around the downstream narrow wall. This configuration produces the highest overall drag coefficient of all the rectangular cavities studied in the present work. With the model at 60° incidence (Fig. 14) the drag remains very high and the pressure distribution is complicated. It is unclear whether the main flow recirculation is parallel to the major or minor axis. By 90° incidence (Fig. 15) the stagnation region has moved to the centre of the downstream wall and the flow is largely symmetric with a dominant vortex structure and a reduced overall drag coefficient.

The variation of overall drag with incidence angle for this depth ratio is illustrated in Fig. 16 which shows both integrated pressure and direct force balance net drag coefficient data normalized by the skin friction coefficient. There is good agreement

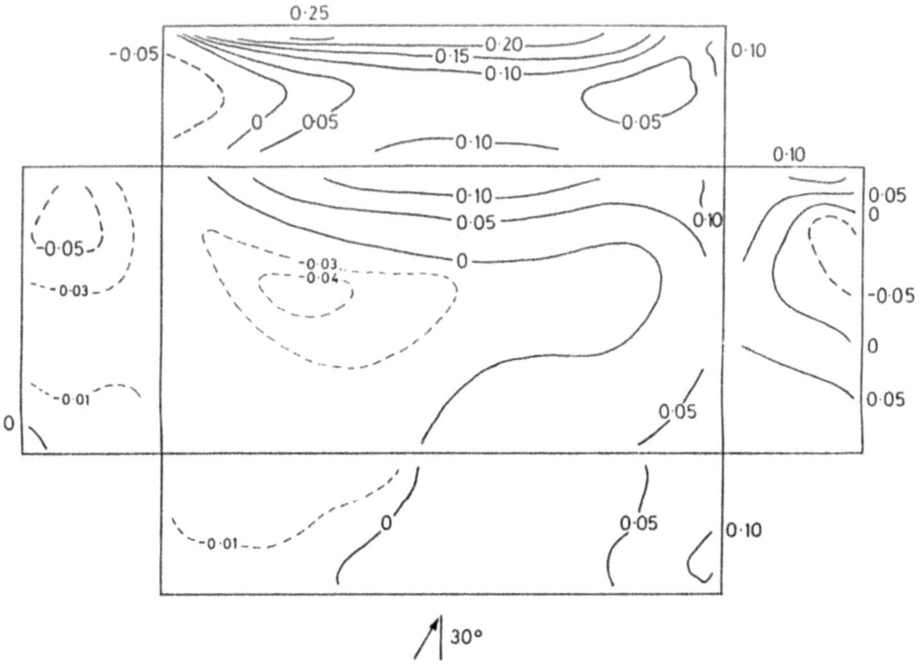

Fig. 12. Pressure distribution with $h/W = 0.5$ for 30° incidence angle.

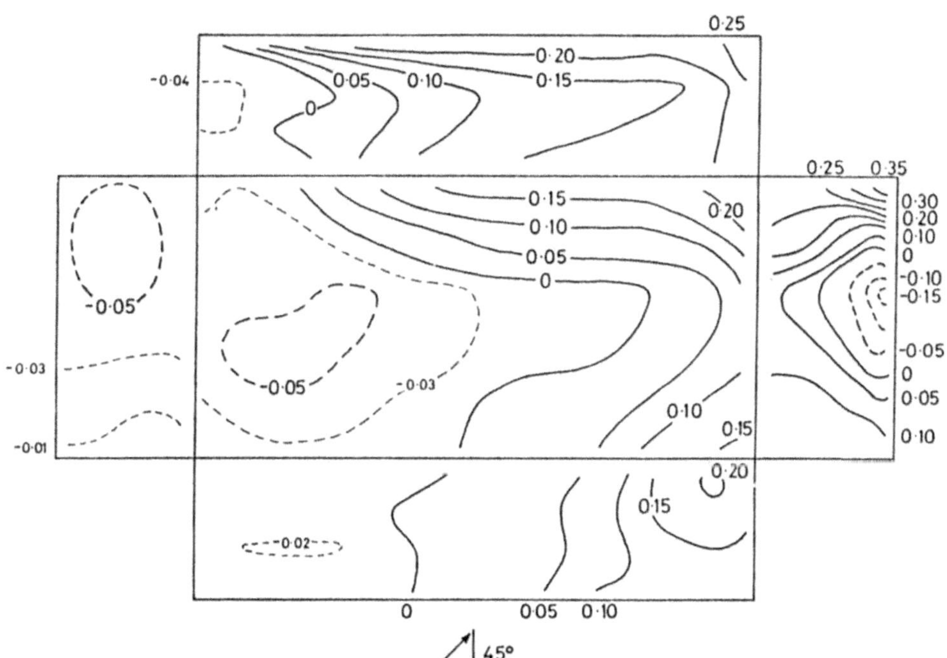

Fig. 13. Pressure distribution with $h/W = 0.5$ for 45° incidence angle.

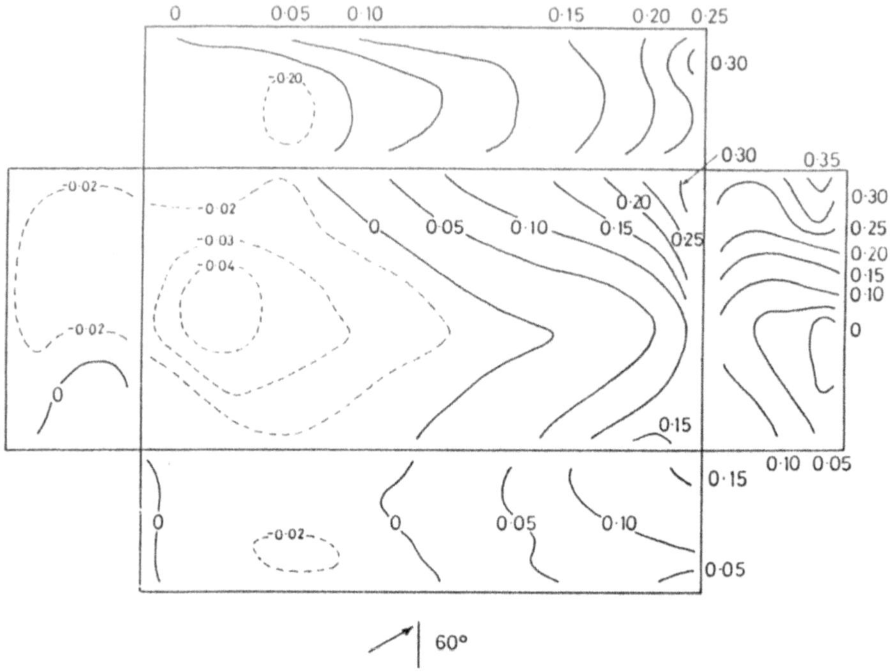

Fig. 14. Pressure distribution with $h/W = 0.5$ for 60° incidence angle.

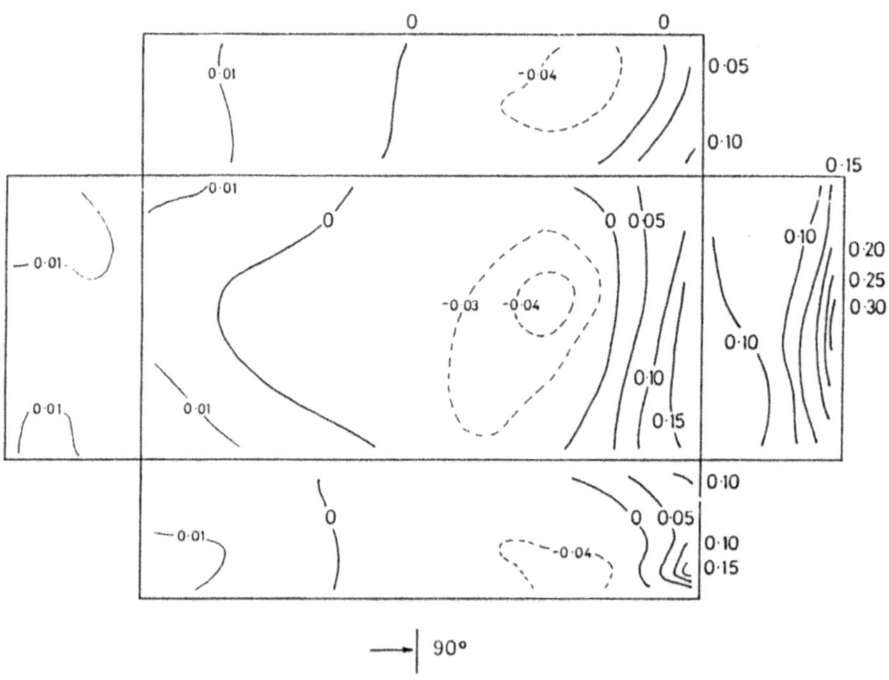

Fig. 15. Pressure distribution with $h/W = 0.5$ for 90° incidence angle.

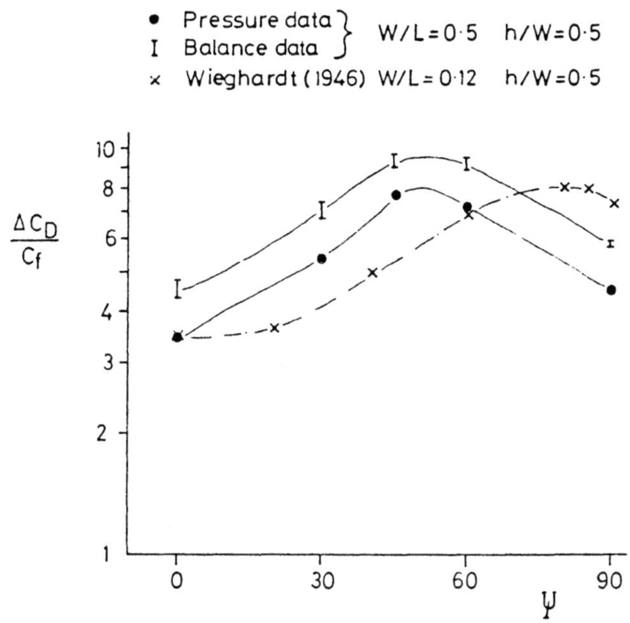

Fig. 16. Variation of normalized drag coefficient with incidence for $h/W = 0.5$

between the trends shown by the two methods which illustrate how the peak drag occurs at an incidence of about 45°. The difference in magnitudes, which manifests itself as a drag offset, is largely due to the fact that the integrated pressure distributions produce only the pressure drag of the cavity whereas the direct balance measurements incorporate the effect of the cavity on the local wall flow regime. The only other available data, that due to Wieghardt (1946), is also shown for comparison, although it should be noted that his results were for a long narrow cavity with W/L of only 0.12. It would appear that the effect of incidence angle on the overall drag coefficient reduces slightly as the cavity becomes more two-dimensional. This is likely to be due to the contribution from the end wall effects, where much of the drag is generated, becoming relatively less as the cavity length is extended.

Rather than attempt to discuss all the remaining distributions individually it is, perhaps, more pertinent to examine the main features in terms of the overall drag and lift coefficients. Figures 17 and 18 show contours of drag and lift coefficients, respectively, in terms of their variation with incidence angle and depth. It should be noted that these values are the integrated pressure data not taking into account the skin friction coefficient. It may be seen that the region of maximum drag occurs for cavities with depth, h/W, of between 0.4 and 0.5 at angles of between 45° and 60°. The lift coefficient data is quite interesting in that for nearly all the cases examined the base was subjected to an overall downward, or thrust, force. Overall lift was only present for some of the deeper cavities. The results show that the thrust is largely insensitive to

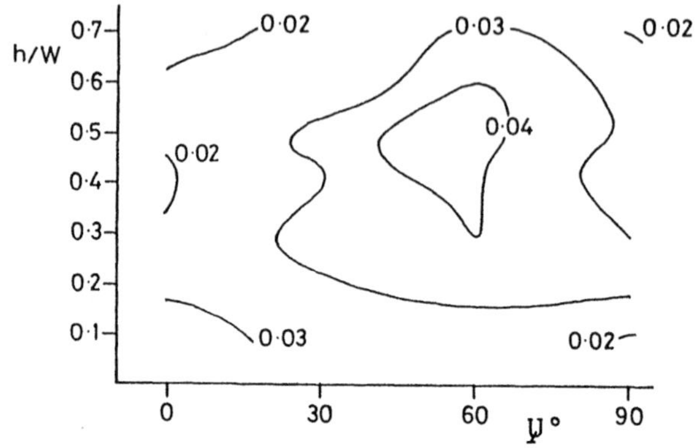

Fig. 17. Variation of drag coefficient with h/W and incidence angle.

changes in incidence angle up to 45° for cavities with h/W less than about 0.3. The larger regions of positive pressure that occur at 90° incidence generally result in higher thrusts at all depths when compared to the other angles.

The variation in drag increment, normalized by the skin friction coefficient with depth for the cavity at 0° incidence is shown in Fig. 19, together with some square planform cavity results and data from other workers. The coefficients shown here are from the pressure measurements. It may be seen that there is a great deal of scatter in both the present data and those from previous research. Nevertheless, there is good agreement between the 0° incidence data ($W/L = 0.5$) and the 90° incidence data ($W/L = 2$ in the notation for this figure) and the general trends are similar to those suggested by the design curve given in AGARD 264, Young and Paterson (1981). In addition, the drag of these rectangular cavities is consistently less than the equivalent square planform cavity. The rapid increase in drag with increase in depth for shallow cavities ($h/W < 0.1$) is reflected in the present data as is the subsequent slower decrease in drag at greater depths. Figure 20 shows the corresponding direct drag balance

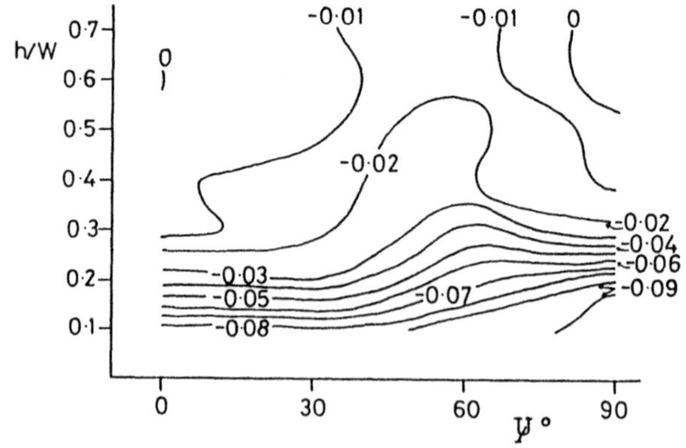

Fig. 18. Variation of lift coefficient with h/W and incidence angle.

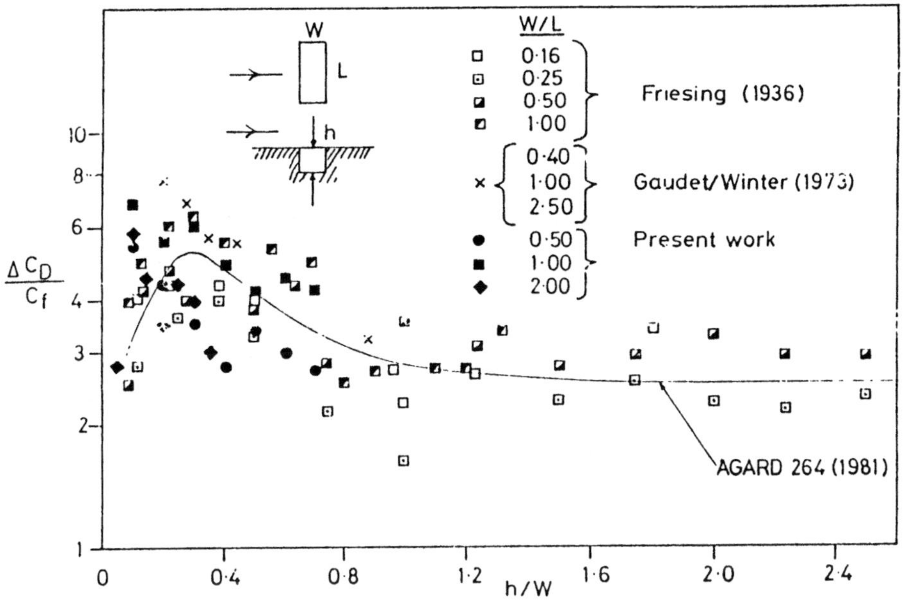

Fig. 19. Variation of normalized drag coefficient with h/W for 0° incidence (pressure data).

measurements which were omitted from the previous figure for clarity. It may be seen that there is good agreement with both the AGARD 264 design curve and with the higher Reynolds number data of Gaudet and Winter (1973). The drag balance results confirm that the rectangular cavity drag coefficient is slightly lower than that for the equivalent square cavity.

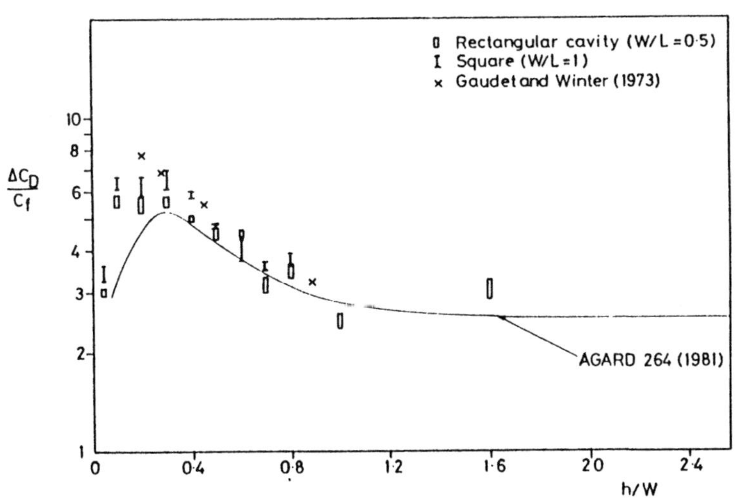

Fig. 20. Variation of normalized drag coefficient with h/W for 0° incidence (balance data).

4. Conclusions

With either the major or minor axis of the rectangular cavities aligned symmetrically to the freestream direction the pressure distribution on the base at all the depth/width ratios studied, showed remarkable lateral uniformity with no noticeable three-dimensionality. No evidence was found to indicate the presence of any cell-like structures within the cavities that had been observed by Maull and East (1963). With the shortest axis parallel to the flow direction the location of the dominating cavity vortex moved upstream and upwards with increasing depth, asymptoting to the location typically observed in two-dimensional rectangular cavity flows. This gradual vortex movement is not so evident when the cavity major axis is aligned parallel to the freestream direction. In terms of the overall net drag increments due to the cavity there is good agreement between the data obtained for these two axis alignments at any given depth/streamwise width ratio. There is also reasonable agreement with previously published data and the design curve given in AGARD 264, Young and Paterson (1981). The drag balance data shows good correlation with the high Reynolds number balance results obtained by Gaudet and Winter (1973).

The effect of yawing the cavity to the flow direction is to create asymmetric flow conditions within the cavity. At depth/width ratios between 0.4 and 0.5 for incidence angles between 45° and 60° this results in high overall drag coefficients. Once again, there is reasonably good agreement between integrated pressure distribution and direct force balance drag measurements. Considering the forces acting on the cavity base, it was found that for nearly all combinations of depth and incidence angle, except for some of the deepest cavities, the base experiences an overall thrust. This force is largely insensitive to changes in incidence angle up to 45° for cavities with $h/W < 0.3$.

The present work has produced a fairly substantial data set which should be of use in the development of improved design guidelines as well as aiding the development of numerical prediction models. Further work is being carried out, concerning the drag of cavities with circular and elliptical planforms, since these configurations exhibit different behaviour from the rectangular case discussed in this paper. Most notably, it is known that for a depth/diameter ratio of about 0.5 the circular cavity flow regime is strongly asymmetric, producing greatly increased drag.

The work described here has concerned cavities in two-dimensional boundary layers. However, in many practical applications the boundary layer will be three-dimensional and there is currently no data available to determine the influence of this three-dimensionality on the cavity flow field. In addition, modifications to the geometries of large rectangular cavities, such as bomb or cargo bays, may well reduce the overall aircraft drag, using techniques such as rounding or chamfering the downstream lip, as summarized by Hoerner (1965). Indeed, this approach is utilized in practice to eliminate unwanted flow oscillations associated with such aircraft cavities, as noted by Rizetta (1988). However, it is unlikely that such modifications to very small cavities would have any noticeable benefits in terms of the total aircraft drag.

Acknowledgements

This work was largely funded by the Department of Trade and Industry (UK) under Agreement No. 2064/120. The authors are indebted to Mr. L. Gaudet of the DRA (Aerospace Division, Bedford) for his support, advice and encouragement throughout the present work. We are also grateful to Mr. T. Markham of British Aerospace, Filton, for supporting the initial proposal regarding the present work. Thanks are also due to the technical staff at both the University of Surrey and the University of Cincinnati.

References

Baysal, O. and Stallings, R.L., Computational and experimental investigation of cavity flow fields. *AIAA J.* 26 (1988) 6–7.

Brandeis, J., Flow separation in shear layer driven cavities. *AIAA J.* 20 (1982) 908–914.

Burggraf, O.R., Analytical and numerical studies of the structure of steady separated flows. *J Fluid Mech.* 24 (1966) 113–151.

Donovan, L.F., A numerical solution of unsteady flow in a two-dimensional square cavity. *AIAA J.* 8 (1970) 524–529.

Friesing, H., Measurement of the drag associated with recessed surfaces: cut-outs of rectangular and elliptical planform. *Z.W.B.F.B.* 628 (1936) (RAE Lib Trans 1614, 1971).

Gaudet, L. and Winter K.G., Measurements of the drag of some characteristic aircraft excrescences immersed in turbulent boundary layers. *RAE Tech Memo Aero* 1538 (1973) (also AGARD CP124, 1973).

Hankey, W.L. and Shang, J.S., Analyses of pressure oscillations in an open cavity. *AIAA J.* 18 (1980) 892–898.

Haugen, R.L. and Dhanak, A.M., Momentum transfer in turbulent separated flow past a rectangular cavity. *J. Applied Mech.* (*Trans ASME E*) 33 (1966) 641–646.

Hoerner, S.F., *Fluid dynamic drag*, pp 5.10–5.11, published by author.

Hunt, J.C.R., Abell, C.J., Peterka, J.A. and Woo, H., Kinematical studies of the flows around free or surface-mounted obstacles applying topology to flow visualisation. *J. Fluid Mech.* 86 (1978) 179–200.

Kaufman, L.G. II, Maciulaitis, A. and Clark, R.L. Mach 0.6 to 3.0 flows over rectangular cavities, Air Force Wright Aero Lab Rept. AFWAL-TR-82-3112, May (1983).

Maull, D.J. and East, L.F., Three-dimensional flow in cavities'. *J. Fluid Mech.* 16 (1963) 620–632.

Mills, R.D., Numerical solutions of the viscous flow equations for a class of closed flows. *J. Royal Aero Soc.* 69 (1965) 714–718.

Nallasamy, M. and Prasad, K.K., On cavity flow at high Reynolds numbers. *J. Fluid Mech.* 79 (1977) 391–414.

Pan, F. and Acrivos, A., Steady flows in rectangular cavities. *J. Fluid Mech.* 28 (1967) 643–655.

Plentovich, E.B., Three-dimensional cavity flows at subsonic and transonic speeds. NASA TM 4209, September (1990)

Rizetta, D.P., Numerical simulation of supersonic flow over a three-dimensional cavity. *AIAA J* 26 (1988) 799–807.

Roshko, A., Some measurements of flow in a rectangular cutout. *NACA TN* 3488, August (1955).

Rossiter, J.E., Wind tunnel experiments on the flow over rectangular cavities at subsonic and transonic speeds. *ARC R & M* 3438, October (1964).

Sinha, S.N., Gupta, A.K. and Oberai, M.M., Laminar separating flow over backsteps and cavities. Part II: Cavities. *AIAA J* 20 (1982) 370–375.

Tani, I, Iuchi, M. and Komoda, H., Experimental investigation of flow separation associated with a step or groove. *Aero Res. Inst. Univ. of Tokyo, Res. Rep. No. 364 119* (1961).

Tillmann, W., Additional measurements of the drag of surface irregularities in turbulent boundary layers. *NACA TM* 1299 (1951) (also: New resistance measurements with surface irregularities in the turbulent boundary layer. *Forschungshefte fur Schiffstechnik* 2 81–88 (BSRA Trans. 322), 1953).
Wieghardt, K., Increase in turbulent frictional resistance caused by surface irregularities, Min. of Air Prod. R&T No. 103, Trans. of FB1563 (1946) (also in Forschungshefte fur Schiffstechnik, 2, 65–71 (BSRA Trans. No. 322, 1953).
Young, A.D. and Paterson, J.H., *Aircraft Excrescence Drag*, NATO AGARD-264, J.L. Jones, ed. (1981).
Zhang, X. and Edwards, J.A., Computational analysis of unsteady supersonic cavity flows driven by thick shear layers. *Aero J.* 92 (1988) 365–374.

Applied Scientific Research **50:** 347–368, 1993.
K. Krishna Prasad (ed.), *Further Developments in Turbulence Management.*

Turbulence structure of dilute polymer and surfactant solutions in artificially roughened pipes

HANS-WERNER BEWERSDORFF[1] & HORST THIEL[2]

[1]*Institute of Hydromechanics and Water Resources Management, Swiss Federal Institute of Technology, CH-8093 Zürich, Switzerland*
[2]*Department of Chemical Engineering, University of Dortmund, P.O. Box 500 500, D-4600 Dortmund 50, Germany*

Abstract. Pressure drop and velocity profile measurements are presented for turbulent flows of drag reducing fluids. The investigation was done in two rough pipes, known as "k"- and "d"-type rough pipes. The results are compared with those obtained in hydraulically smooth pipe of identical diameter. The spatial arrangement of the roughness elements in the pipe determines the parallel shift in the elastic sublayer and in the core region of the dimensionless turbulent velocity profile. The slopes of the velocity profiles in these regions remain unaffected by the arrangement which is an indication that the hydrodynamic influence of the roughness is restricted to the near-wall region. The drag reducing surfactant solution exhibited a drag reduction in the smooth as well as in the rough pipes which was higher than that given by Virk's maximum drag reduction asymptote. For this solution no influence of the roughness on the turbulence was detected when the dimensionless roughness height in viscous units was less than 12.

1. Introduction

Drag reduction obtained by the addition of soluble polymers or surfactants is well known for turbulent flows in smooth pipes. The state of the art is well documented in review articles [1–4] and recent conference proceedings [5–6]. However, only a few publications describe drag reduction in rough pipes. Lindgren and Hoot [7] first detected drag reduction for a 60 ppm Polyox solution in a rectangular channel with rough walls.

Due to the manufacturing process commercial (steel) pipes normally exhibit a certain roughness which is important for many industrial applications of drag reduction [8].

Although the mechanism of the interaction between the additives and the turbulence is still unknown it is generally accepted that the non-Newtonian rheological properties of the polymer or surfactant solution are responsible for the occurrence of drag reduction. Drag reducing polymer or surfactant solutions are viscoelastic fluids. In elongational fields, which exist in turbulent flows, the elongational viscosity of dilute polymer solutions can be increased considerably in comparison to that of the Newtonian solvent. This results in an additional resistance against vortex stretching and leads to a reduction of the energy dissipation.

The drag reduction in turbulent flows of polymer and surfactant solutions is accompanied by a change in the structure of turbulence, which can depend on the type of additive. In homogeneous dilute polymer solutions a so-called elastic sublayer forms in the dimensionless velocity profile between the viscous sublayer and the turbulent core. This elastic sublayer increases in thickness with increasing drag reduction and at maximum drag reduction it reaches to the centre of the pipe or channel. No turbulent core then exists, and the friction behaviour is given by Virk's [1] empirical maximum drag reduction asymptote

$$f^{-1/2} = 19\log(Re f^{1/2}) - 32.4. \tag{1}$$

Here, f means the Fanning friction factor defined by

$$f = \frac{2\tau_w}{\rho U^2}, \tag{2}$$

where τ_w is the wall shear stress, ρ the density of the fluid, and U the bulk velocity. The Reynolds number used in Eq. (1) is defined as

$$Re = \frac{Ud}{\nu}, \tag{3}$$

for a pipe flow, where d is the pipe diameter and ν the kinematic viscosity of the fluid.

In pipe and channel flows of drag reducing fluids the turbulence intensities are also affected. The maximum of the axial turbulence intensity is shifted to larger wall distances [9, 10], whereas the maximum of the radial turbulence intensity, which occurs in turbulent flows of Newtonian fluids at $y^+ = 50 - 70$, disappears [9, 10], i.e. the radial turbulence intensity now reaches its maximum in the centre of the pipe. Simultaneously a strong decrease of the radial turbulence intensity occurs [9, 10]. In the case of maximum drag reduction conditions according to Eq. (1), the axial turbulence intensity is also decreased [10]. Similarily the Reynolds shear stresses near the wall are also lower than in the Newtonian solvent [10–12]. In turbulent flows of drag reducing fluids the burst-sweep mechanism is modified. This mechanism plays a major role in the generation of the Reynolds shear stresses and is therefore important for the turbulent momentum transfer. It has been shown [13] that the burst frequency and intensity are affected in the case of drag reduction.

Previous studies on drag reduction in rough pipes mainly focus on the friction behaviour of polymer solutions. An extensive study on the friction behaviour of homogeneous polymer solutions in sand-roughened pipes was undertaken by Virk [14]. He used polymers of different molecular weights, and varied the polymer concentration, and the relative roughness d/k of the pipe, i.e. the ratio of the pipe diameter to the roughness height. The maximum drag reduction according to Eq. (1) could also be reached in rough pipes provided that the dimensionless roughness k^+ was less than 12, where

$$k^+ = \frac{k u_\tau}{\nu}. \tag{4}$$

In this equation u_τ is the friction velocity defined by

$$u_\tau = (\tau_w/\rho)^{1/2}. \tag{5}$$

In other words, at maximum drag reduction conditions the friction behaviour in rough pipe is identical to that in a hydraulically smooth pipe for $k^+ < 12$. When k^+ exceeds 12 the friction factor in rough pipes is larger than in smooth ones. Furthermore, Virk [12] showed that the onset of drag reduction occurred at the same wall shear stress in smooth and rough pipes. Up to a dimensionless roughness $k^+ = 50$, the drag reduction was identical in a smooth and a rough pipe at the same friction velocity u_τ. For $k^+ > 50$ the drag reduction in a rough pipe was lower than that found in a smooth pipe.

In general, the drag reduction in rough pipes exhibits a maximum at a certain Reynolds number. McNally [15] detected that the maximum drag reduction always occurred at the same friction velocity regardless of the polymer concentration. The decrease in drag reduction at high Reynolds numbers beyond the maximum was related to mechanical degradation of the polymers due to the influence of the roughness by Debrule and Sabersky [16], whereas Spangler [17] and Virk [14] found hydrodynamic effects to be responsible for the decrease. Spangler [17] mounted a smooth pipe downstream of the rough pipe, and measurements in the smooth pipe demonstrated that no degradation occurred. In the experiments of Virk [14] the pressure drop at Reynolds numbers above the maximum drag reduction remained nearly constant at a constant flow-rate in a closed loop after several runs, and did not increase as one would expect in case of mechanical degradation. It is the aim of the present investigation to study the influence of artificial roughnesses on the turbulence structure in dilute polymer and surfactant solutions.

2. Experimental

In experimental studies on the influence of roughnesses mainly artificially produced roughnesses are used. The advantage of this is that in this way the geometry of the artificial roughness elements is well-defined which allows a characterization of the influence of the roughness on the hydrodynamic behaviour. In commercial rough pipes a nonuniform distribution of roughness height and location of roughness elements on the pipe surface exists.

In the present study two commercially available PVC-pipes of an outer diameter of 63 mm with a wall thickness of 7 mm and a length of 16 m were used. On the inner surface of the pipes rectangular grooves were cut around the circumference normal to the flow direction. The spacing of the grooves in the flow direction was constant for each pipe. The dimensions of the roughness elements were identical in both pipes, however, the spacing of the roughness elements differed. The dimensions and the spacing of the roughness elements is shown in Fig. 1.

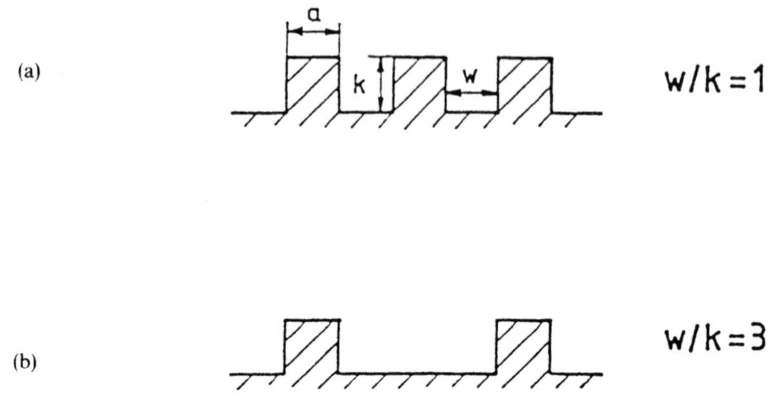

Fig. 1. Arrangements of the roughness elements. (a) "d"-type roughness; (b) "k"-type roughness.

The two arrangements of the roughness elements are essentially identical to those of Perry, Schofield and Joubert [18] who found that the different arrangements influenced a turbulent boundary layer in a different way. The terms "d"-type and "k"-type roughness are adopted in the present work. The dimensions of the two rough pipes are listed in Table 1. The average diameter d was calculated according to

$$d = \left(\frac{4V}{\pi L}\right)^{1/2}. \tag{6}$$

In order to do this the content in liters of a pipe segment of a length L was measured.

Due to manufacturing reasons only rough pipes of a length of approximately 0.3 m could be produced. Four of these pipes were glued together. By clamping 15 of these segments together, the total length of the test pipes became approximately 16 m. An exact check of the manufacturing process ensured that no nonuniformity occurred in the periodic spacing of the roughness elements. In addition, a hydraulically smooth perspex pipe of 16 m length and an internal diameter of 50 mm was used in order to compare the results for the rough pipes with those for a smooth pipe.

The experimental set-up is shown in Fig. 2. The water, the polymer or the surfactant solutions were pumped from a storage tank by moyno pumps. The temperature was recorded by a resistance thermometer. The flow rate was measured by an inductive

Table 1. Dimensions of the rough pipes.

Surface	a [mm]	k [mm]	w [mm]	d [mm]
"k"-type roughness	1	1	3	50.5
"d"-type roughness	1	1	1	50

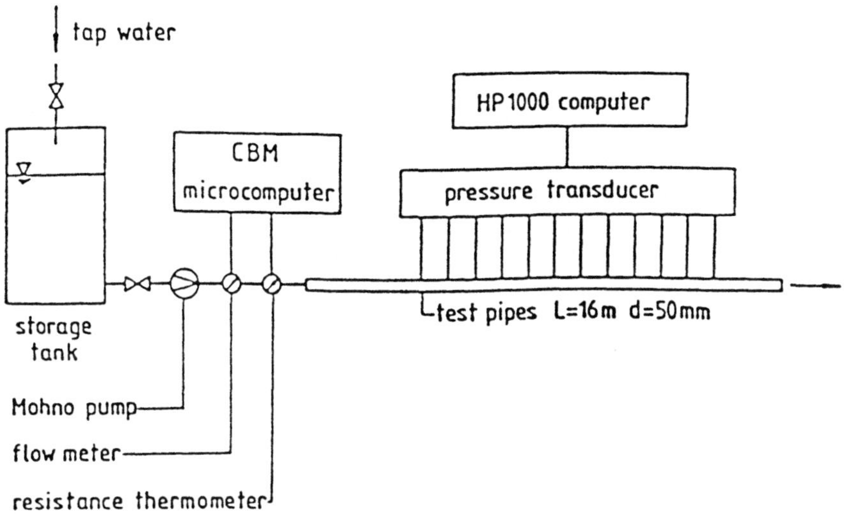

Fig. 2. Experimental arrangement.

flowmeter (Fischer & Porter, model D10 DK1425) with an accuracy of ±0.5%. For proper operation this inductive flowmeter requires a minimum electrical conductivity of 5 μS/cm, a condition which was fulfilled in all experiments, because tap water or solutions of polymers or surfactants in tap water were used as test fluids. In order to determine the Reynolds number the two signals of the resistance thermometer and the inductive flowmeter were fed into a microcomputer. To measure the pressure drop, holes were drilled into the pipes. They were spaced at intervals of 100 cm in the smooth pipe and at 110 cm in the two rough pipes. According to Nunner [19] and Möbius [20] these holes were drilled in the centre between the adjacent roughness elements in order to avoid errors in the pressure drop experiments. The pressure drop between two taps was measured by differential pressure transducers (Philipps, model PD2 and Siemens, model Teleperm F). A connection of the outputs of the pressure transducers to a computer system enabled a direct analysis of the experimental data.

15 m downstream of the pipe inlet velocity profiles were measured by a DISA Laser-Doppler anemometer (LDA-module-optics 55x) working in the backscattering mode. The dimensions of the measuring volume were 0.5 mm in the long axis and 0.05 mm in the short axis with a fringe spacing of 5 μm. In order to improve the quality of the Doppler-signal aluminium particles (particle size 3–10 μm) were added at a concentration of 5–10 ppm. The Doppler-signals were sampled with a frequency of 100 Hz and fed into a computer system. The local averaged velocities result from averaging 5000 data points. By increasing the number of data points it was checked, that the mean values of the velocity did not change.

The polymer used was polyacrylamide, SEPARAN AP 45 (tradename from Dow Chemical Company). According to the manufacturer [21] its molecular weight is 4–$5 \cdot 10^6$ g/mol. The polymer powder was first suspended in isopropanol and then added to the desired amount of water. The solutions were stirred at low speed for 24 hours.

The surfactant solutions were produced by dissolving equimolar amounts of tetradecyltrimethylammoniumbromide (C_{14}TABr) and sodiumsalicylate (NaSal). In this way the surfactant tetradecyltrimethylammoniumsalicylate (C_{14}TASal) with an equimolar amount of sodiumbromide (NaBr) is obtained. Above a critical concentration the surfactant molecules assemble and build-up rod-like micelles which are necessary [22] to achieve drag reduction in turbulent flows.

3. Experimental results

As mentioned in Chapter 2 the two types of roughness elements interact with the near-wall turbulence in a different way. In a "k"-type roughness an extensive exchange of fluid between the grooves and the outer flow takes place, whereas in a "d"-type roughness stable vortices exist inside the grooves and the outer fluid flows nearly undisturbed over the top of the roughness elements [18]. Therefore it is expected that these two types of roughnesses exhibit a different effect on the friction behaviour and on the turbulent dimensionless velocity profiles. In Fig. 3 the friction behaviour of the two rough pipes is shown for water. For the smooth pipe the deviations from the

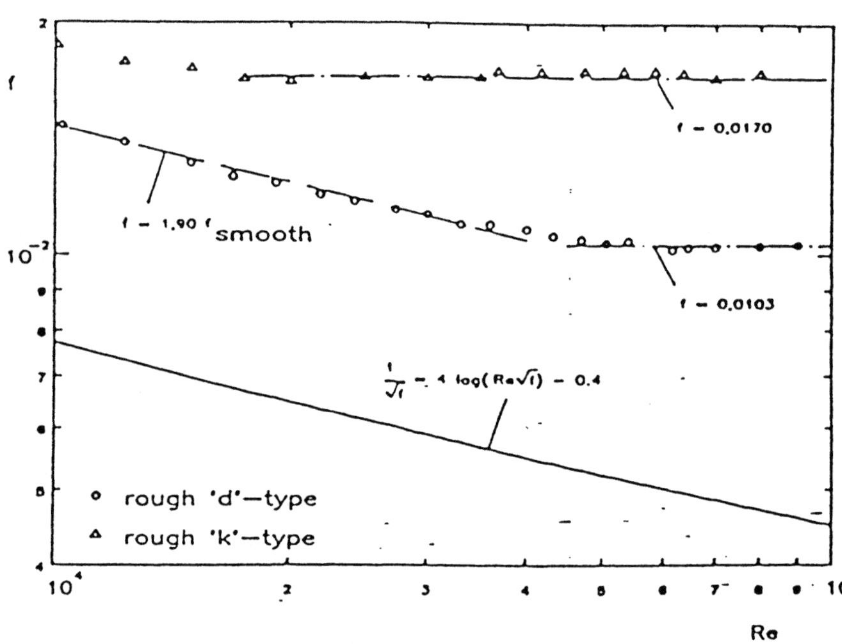

Fig. 3. Friction behaviour for a water flow in the rough pipes.

Prandtl-Kármán law

$$f^{-1/2} = 4\log(Ref^{1/2}) - 0.4, \tag{7}$$

were. found to be less than three percent in the Reynolds number range $15{,}000 < Re < 90{,}000$. The two types of artificial roughness have a different influence on the value of the friction factor and its dependence on the Reynolds number. The friction factors for the "k"-type roughness do not depend on the Reynolds number for $Re > 15{,}000$, i.e. the fully rough flow regime is reached. This type of friction behaviour can also be obtained for the "d"-type of roughness for $Re > 40{,}000$, but the friction factors are lower than those of the "k"-type roughness. In the Reynolds number interval $10{,}000 < Re < 40{,}000$ the friction behaviour for the "d"-type roughness is parallel to the Prandtl-Kármán law, Eq. (7). For $Re > 50{,}000$ the friction factors for the "d"-type rough pipe become also constant regardless of the Reynolds number which indicates that the fully rough flow regime is reached. In addition, it elucidates that the "d"-type rough pipe becomes a "k"-type rough pipe at high Reynolds numbers. Consequently, it shows that the type of roughness depends on w/k and on the Reynolds number.

According to Hopf [23] a friction behaviour which can be expressed by

$$f = \xi f_{\text{smooth}} \tag{8}$$

is characteristic for "wall ripples" where f_{smooth} means the friction factor in a smooth pipe. Morris [24] developed a similar friction law for so-called "quasi-smooth" flows.

In general, it is convenient to convert a given roughness to an equivalent sand roughness by inserting the measured friction factors in the empirical friction law of Nikuradse [25] for sand-type roughnesses in the fully rough flow regime

$$f^{-1/2} = 2.28 + 4\log(d/k_s). \tag{9}$$

Equivalent sand roughnesses are listed in handbooks for commercial rough pipes of different materials. This simplifies the calculation of the pressure drop in pipelines in practical situations because convenient relationships, such as the Colebrook-White equation [26] or Eq. (9), can be used.

This normalization procedure allows for comparisons of measurements with published data for artificially rough pipes. By using Eq. (9) for the data of the fully rough flow regime, the "k"-type roughness gives an equivalent sand roughness of $k_s = 2.4\,\text{mm}$ ($k_s/k = 2.4$) and the "d"-type roughness gives $k_s = 0.64\,\text{mm}$ ($k_s/k = 0.64$).

Turbulent velocity profiles are normally presented in terms of wall variables with the nondimensional velocity

$$U^+ = \frac{\bar{u}}{u_\tau} \tag{10}$$

plotted versus the logarithm of a dimensionless wall distance

$$y^{+} = \frac{yu_{\tau}}{\nu} \tag{11}$$

where $\bar{u}$ is the time-averaged local velocity and y the wall distance. According to Schlichting [27] turbulent velocity profiles of Newtonian fluids in a smooth pipe can be divided into three regions. Close to the wall ($y^{+} < 5$) a viscous sublayer exists with

$$U^{+} = y^{+}. \tag{12}$$

In this region viscous forces are dominating. With increasing wall distance the Reynolds stresses become important and in the buffer layer ($5 < y^{+} < 30$) the Reynolds stresses are of the same order of magnitude as the viscous stresses. For $y^{+} > 30$ the Reynolds stresses are dominating and the velocity profile is given by

$$U^{+} = A \ln y^{+} + B \tag{13}$$

where the universal constants A and B arc determined from experiments to be $A = 2.5$ and $B = 5.5$. Figure 4 shows dimensionless velocity profiles of a turbulent flow of water in a smooth pipe at two Reynolds numbers.

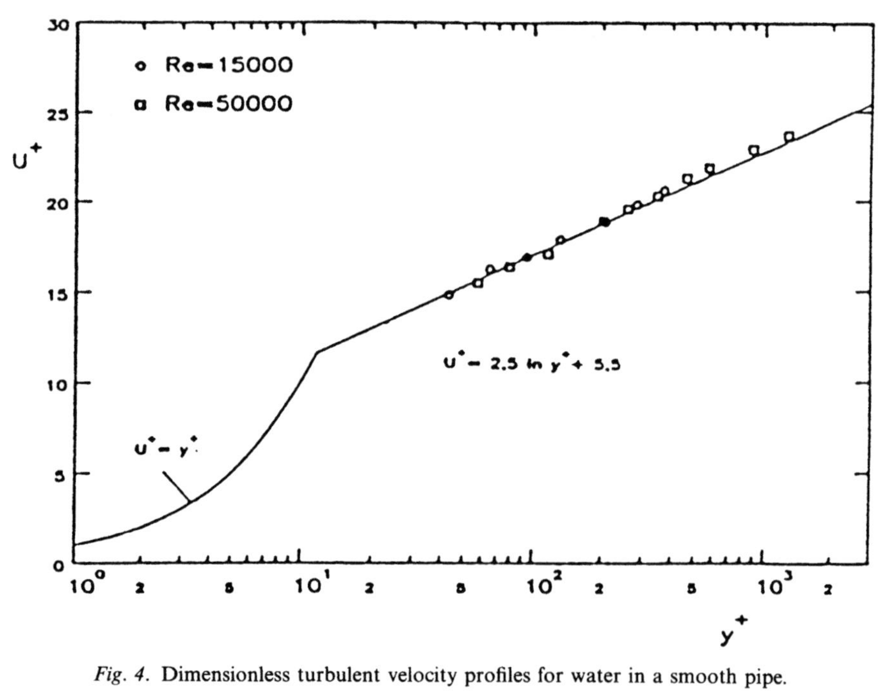

Fig. 4. Dimensionless turbulent velocity profiles for water in a smooth pipe.

Due to the signal/noise ratio of the Doppler-signal the data for the near-wall region have been omitted. For $y^+ > 40$ the data exhibit only small deviations from Eq. (13) with the constants given above, which shows the accuracy of the LDA-measurements.

In turbulent flows over a rough wall the logarithmic region is shifted parallel to the one for smooth walls but the dimensionless velocities are lower in comparison to Eq. (13). Consequently, this logarithmic region can be described by

$$U^+ = A \ln y^+ + B - \Delta u/u_\tau \tag{14}$$

with the constants A and B given above [28]. $\Delta u/u_\tau$ is called the roughness function. Besides the magnitude of the parallel shift of the logarithmic layer it provides, according to Bandhyopadhyay [29], the increased friction due to the surface roughness in comparison to a smooth wall.

In a study with turbulent flows over extreme roughnesses Einstein and El-Samni [30] found that a logarithmic relationship according to Eq. (14) only exist when the origin of the velocity profile, i.e. the location where the wall distance $y = 0$, is located between the top and the foot of the roughness element. Clauser [31] proposed a method for estimating the origin of turbulent velocity profiles over roughness elements (see Fig. 5). The dimensionless velocities U^+ are plotted as a function of the logarithm of the wall distance

$$y = y_T + e. \tag{15}$$

Different values of e influence the slope of the logarithmic region of the velocity profile. In order to normalize the velocity profiles the value of e is generally adopted which results in a slope of $A = 2.5$ in the logarithmic layer. The literature provides values for e for "k"-type roughnesses with $w/k = 3$ as $e = 0.7\,k$ [32] and $e = 0.8\,k$ [33]. In the present investigation a value of the "k"-type roughness a value of $e = 0.75\,k$ was used to normalize the velocity profiles. For the "*d*"-type roughness a value of $e = 0.5\,k$ is used [29].

Dimensionless velocity profiles for turbulent flows of water over the two types of roughnesses are shown in Figs 6 and 7.

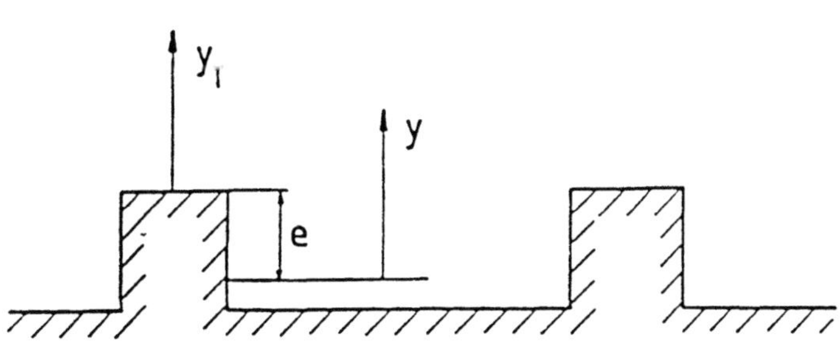

Fig. 5. Origin of the wall distance on rough surfaces.

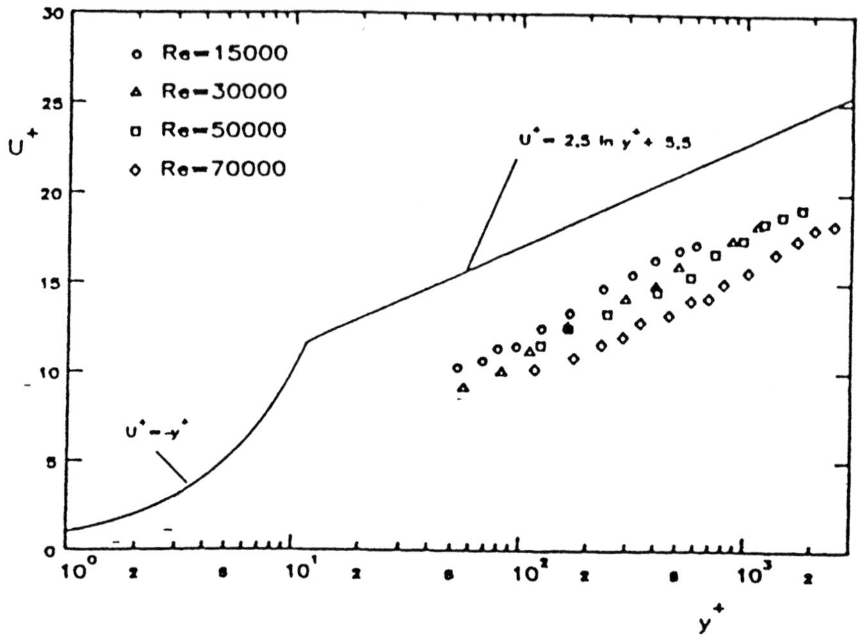

Fig. 6. Dimensionless velocity profile of a turbulent flow of water in the "d"-type rough pipe.

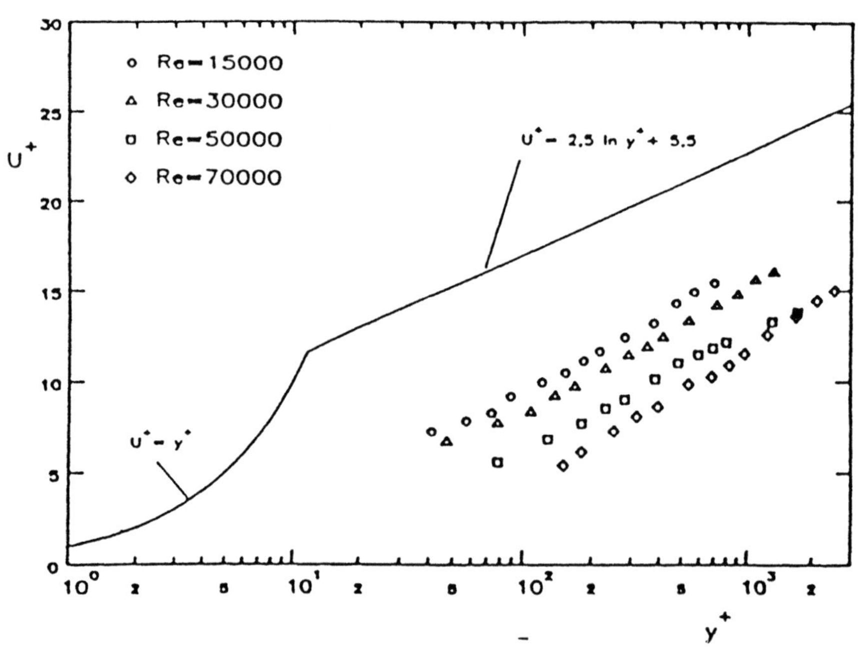

Fig. 7. Dimensionless velocity profile of a turbulent flow of water in the "k"-type rough pipe.

Depending on the type of roughness and the Reynolds number the logarithmic layers have been shifted to lower dimensionless velocities in comparison to a smooth pipe. Table 2 provides the constants A and B according to Eq. (13) and the dimensionless roughness k^+ according to Eq. (4). Using the constants A and B of Table 2 the roughness function $\Delta u/u_\tau$ can be calculated according to Eq. (14) from the measured velocity profiles. Furthermore, it also can be calculated from the friction behaviour to

$$\Delta u/u_\tau = (2/f_{\text{smooth}})^{1/2} - (2/f_{\text{rough}})^{1/2} \tag{16}$$

where f_{smooth} and f_{rough} have to be taken at identical friction velocities [29]. Figure 8 shows the calculated roughness functions as a function of the dimensionless roughness.

The band width of the data of the roughness function originating from the velocity profiles results from the deviations in the slope of the velocity profiles from $A = 2.5$ (see Table 2). A comparison of these data with those calculated from the friction behaviour by using Eq. (16) gives a rather good agreement. At the same dimensionless roughness the roughness function for the "k"-type roughness is generally larger than that of the "d"-type roughness.

Perry and Joubert [33] give for $w/k = 3$ the equation

$$\Delta u/u_\tau = 5.6 \log k^+ - 0.2 \tag{17}$$

for $100 < k^+ < 700$ which is based on their own measurements and on the data of Moore [32]. The data resulting from this equation are added as a solid line in Fig. 8. The deviations from the measured data of the present work are small. For a "d"-type roughness equations similar to Eq. (17) do not exist.

Table 2. Constants A and B of the velocity profile and dimensionless roughness height

Surface	*Re*	*A*	*B*	k^+
Smooth	15,000	2.7	4.3	—
	50,000	2.8	4.1	—
"d"-type rough	15,000	2.9	−1.2	24
	30,000	3	−3.2	48
	50,000	2.9	−2.7	77
	70,000	2.8	−3.6	106
"k"-type rough	15,000	3	−4.2	29
	30,000	2.9	−5	54
	50,000	2.8	−6.5	90
	70,000	3.3	−11.2	126

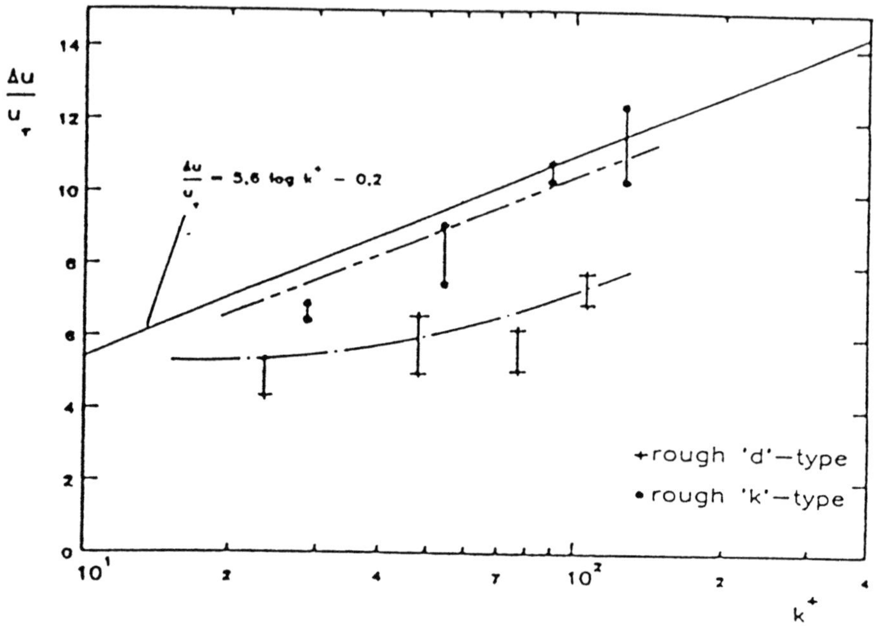

Fig. 8. Roughness function as a function of the dimensionless roughness data points: calculated from the measured velocity profiles by using Eq. (14) broken lines: calculated from the friction behaviour by using Eq. (16).

The friction behaviour of a solution of Separan AP45 at $c = 20\,\mathrm{ppm}$ in the two rough pipes and in the smooth pipe is shown in Fig. 9.

The different surface finish of the test pipes influences the wall shear stress at onset conditions as well as the magnitude of drag reduction. When the friction curves of the polymer solution in the different pipes are extrapolated to those of the solvent it becomes evident that the onset is shifted to higher wall shear stresses with increasing influence of the roughness ($Re\{f\}^{1/2}_{\text{smooth}} = 270$; $Re\{f\}^{1/2}_{\text{“d”-type}} = 1{,}720$; $Re\{f\}^{1/2}_{\text{“k”-type}} = 2{,}500$). This is accompanied by a decrease of the magnitude of attainable drag reduction DR (see Fig. 10), which is defined as $\mathrm{DR} = 1 - f_S/f_N$ where f_S is the friction factor of the polymer or surfactant solution and f_N the friction factor of the Newtonian solvent, both measured at the same flow-rate.

In the “k”-type rough pipe the drag reduction appears to reach a constant value at high friction velocities which would be consistent with a fully rough flow regime. This trend was also observed in [34]. A lower drag reduction in rough pipes in comparison to smooth pipes was also observed by McNally [15] and Ivanyuta and Chekalova [35]. However, this is in contrast to the findings of [14] and [34] where no influence of the surface finish on the magnitude of drag reduction and the onset-point was detected provided that k_s^+ is lower than about 50. Possible explanations for these discrepancies may be the type of roughness or the smaller diameters in the earlier studies ($d = 8.5\,\mathrm{mm}$ in [14] and $d = 30.7\,\mathrm{mm}$ in [34]).

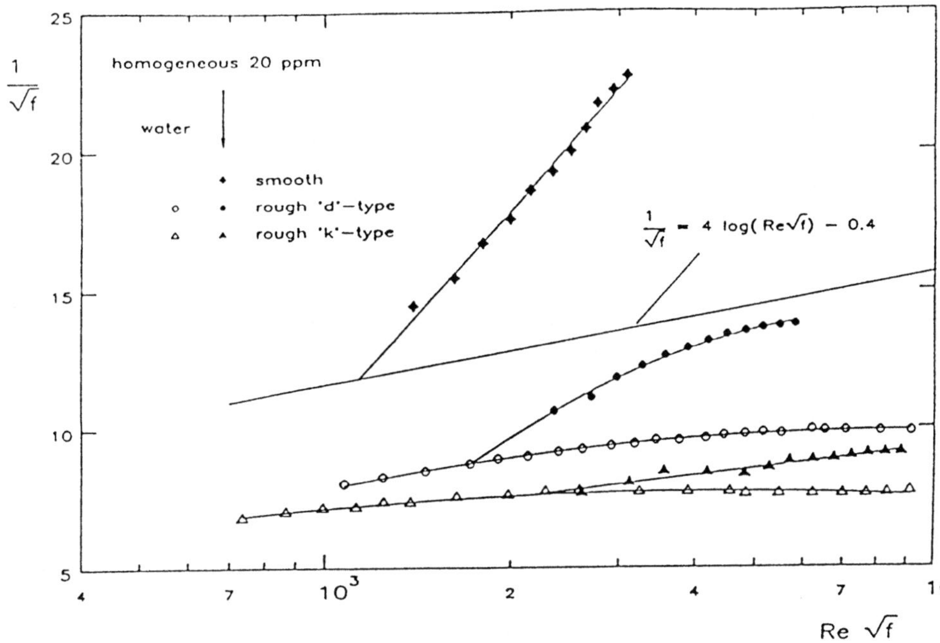

Fig. 9. Friction behaviour of a homogeneous polymer solution at $c = 20\,\text{ppm}$ in Prandtl-Kármán coordinates.

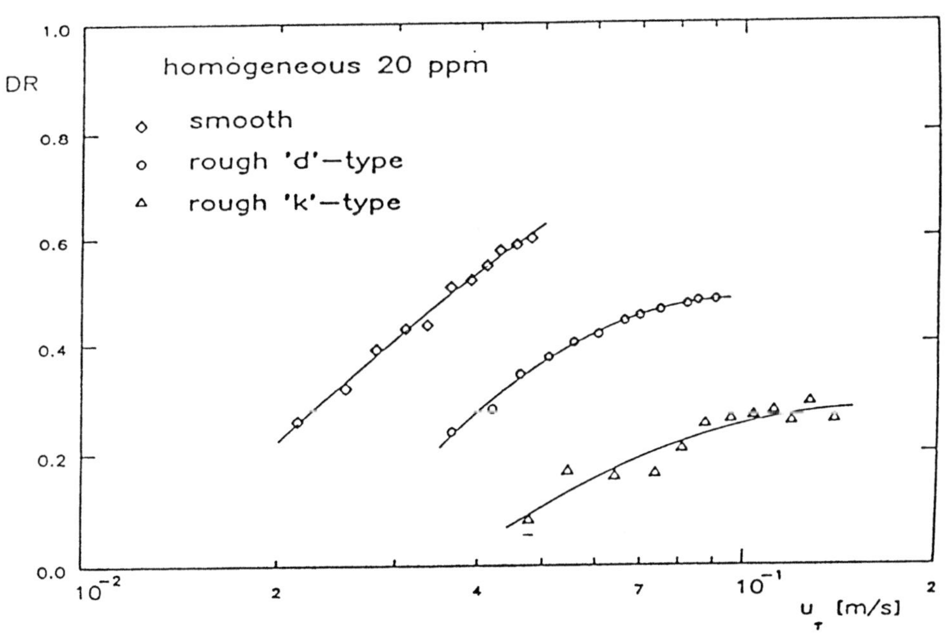

Fig. 10. Drag reduction as a function of the friction velocity for a homogeneous polymer solution.

The drag reduction in homogeneous polymer solutions is accompanied by a change in the shape of the turbulent velocity profiles. For turbulent velocity profiles in homogeneous dilute drag reducing polymer solutions the three-layer-model of Virk [1] gives satisfactory results. The viscous sublayer near the wall, Eq. (12), seems to remain unchanged in comparison to a Newtonian fluid. Further away from the wall instead of the buffer layer, a new layer, a so-called elastic sublayer, forms, in which

$$U^+ = 11.7 \ln y^+ - 17 \tag{18}$$

which is also called Virk's [1] ultimate profile. In the core region the velocity profile exhibits a parallel shift to higher dimensionless velocities

$$U^+ = 2.5 \ln y^+ + 5.5 + \Delta B. \tag{19}$$

Hendricks et al. [36] first published velocity profile measurements of a drag reducing polymer solution in a rough channel flow. Their results are replotted in Fig. 11.

In the dimensionless velocity profile two regions can be distinguished. In the elastic sublayer their velocity data in the hydrodynamic smooth channel are well represented by Virk's ultimate profile, Eq. (18), while Eq. (19) applies in the core region. In the

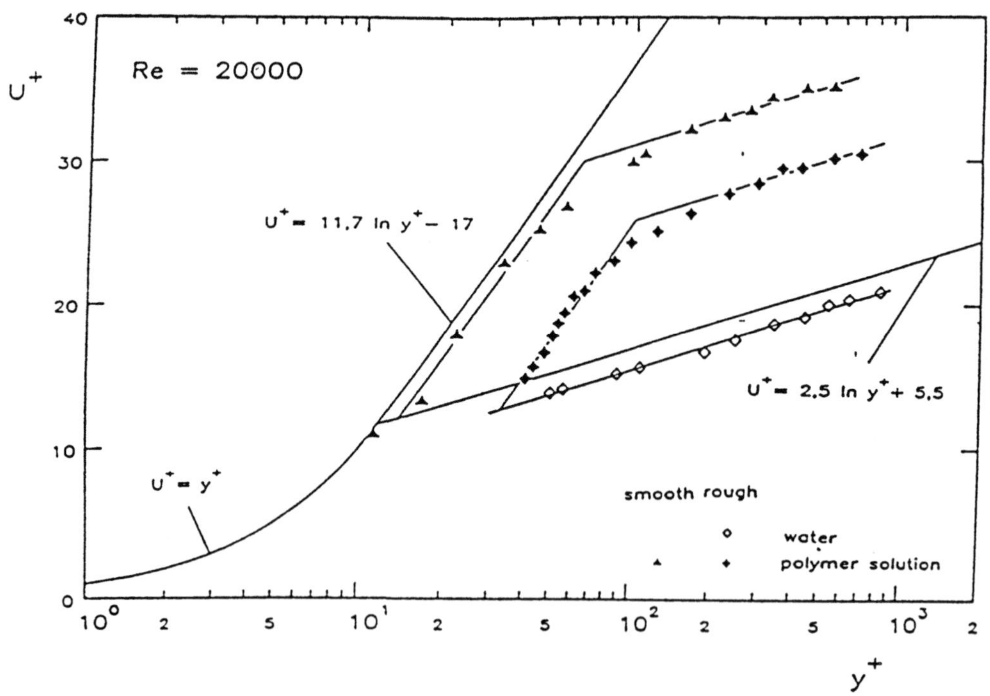

Fig. 11. Dimensionless turbulent velocity profiles of a polymer solution in a smooth and in a rough channel flow (60 ppm Polyox WSR 301, from [36]).

rough channel flow both regions of the velocity profile have been shifted to lower dimensionless velocity values. The data in the elastic sublayer follow

$$U^+ = 11.7 \ln y^+ - 17 + \Delta u_e/u_\tau \tag{20}$$

and in the turbulent core region

$$U^+ = 2.5 \ln y^+ + 5.5 + \Delta B - \Delta u/u_\tau. \tag{21}$$

The roughness functions $\Delta u_e/u_\tau$ and $\Delta u/u_\tau$ describe the parallel shift of the profiles in smooth pipes [Eqs. (18) and (19)] due to the hydrodynamic influence of the roughness. Figures 12 and 13 show the velocity profile measured in the smooth and in the two rough pipes at two Reynolds numbers. Due to the poor quality of the Doppler signal in the near-wall region it was only possible to obtain velocity data in the core region. Therefore, the existence of an elastic sublayer according to Eq. (20) could not be verified experimentally. In comparison to the Newtonian water data the dimensionless velocity data of the drag reducing polymer solution are shifted to higher dimensionless velocities. The data from the smooth pipe can be approximated by Eq. (21). The magnitude of drag reduction, which decreases from smooth via "d"-type rough to "k"-type rough, is reflected in the magnitude of ΔB.

Before turning to the results of the drag reducing surfactant solution it is essential to consider the rheological behaviour of these solutions which differs from that of drag

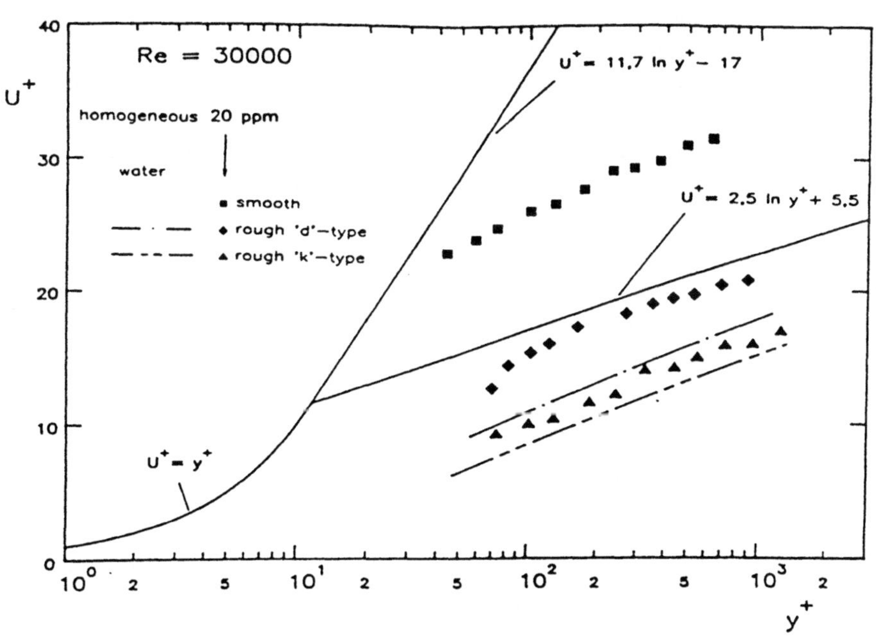

Fig. 12. Dimensionless velocity profiles of a homogeneous dilute polymer solution at $Re = 30{,}000$.

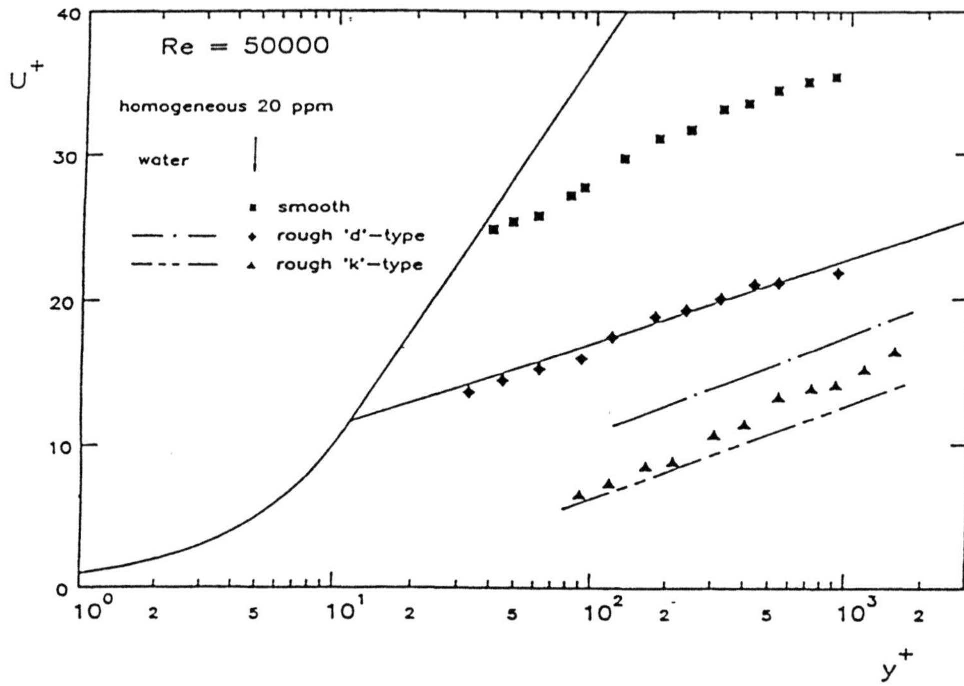

Fig. 13. Dimensionless velocity profile of a homogeneous dilute polymer solution at $Re = 50{,}000$.

reducing polymer solutions. In the experiments a C_{14}TASal-solution (see section 2) was used as a test fluid at a concentration of $c = 945$ ppm. Figure 14 shows the behaviour of the shear viscosity η as a function of the shear rate D for such a solution. The measurements were done with a Bohlin VOR-rheometer (measuring system CS). In the shear rate range $D < 100\,\mathrm{s}^{-1}$ the solution behaves as a Newtonian fluid. Due to the concentration of dissolved surfactants the shear viscosity is slightly increased in comparison to the solvent.

In the shear rate range $100\,\mathrm{s}^{-1} < D < 300\,\mathrm{s}^{-1}$ the shear viscosity increases suddenly. This phenomenon is called shear induced state, SIS. In [22] this behaviour is explained by the formation of superordered structures which consist of single rod-like micelles. The shear rate at which the SIS occurs depends on the surfactant concentration, any additional electrolyte concentration, the temperature, and, surprisingly, on the geometry of the viscometer [22, 37], e.g. on the gap width of a Couette viscometer [37, 38]. This is also an indication that the size of the superordered structures in the SIS is of the order of magnitude of the gap width [37, 38].

In a rheological study on the behaviour of drag reducing surfactant solutions in elongational fields [39] an increase of the elongational viscosity in comparison to that of the solvent was detected. This increase occurred analogously to the increase of the shear viscosity in the SIS. When the Trouton ratio (elongational/shear viscosity) was

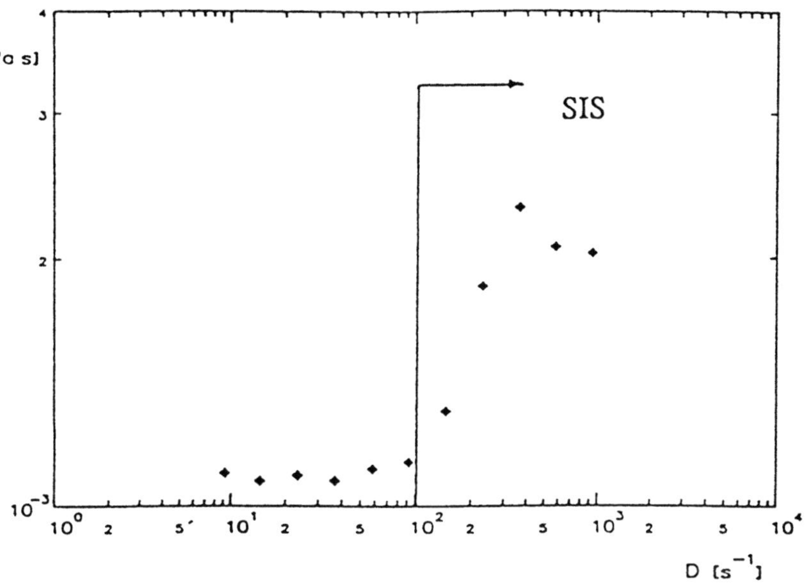

Fig. 14. Shear viscosity of C_{14}TASal-solution at $c = 945$ ppm and $T = 21°C$.

calculated a value of three was obtained in the whole range of studied strain rates. In the same study this Trouton ratio was increased for drag reducing polymer solutions. Therefore it seems that the increase in elongational viscosity of drag reducing surfactant solutions plays a minor role for the mechanism of drag reduction.

One of the advantages of using surfactants as drag reducers instead of polymers is that a high drag reduction can also be obtained in large pipes. Figure 15 shows the friction behaviour of the C_{14}TASal-solution at $c = 945$ ppm in the three different pipes. The Reynolds number is based on the solvent viscosity.

In all pipes the surfactant solution exhibits drag reduction, and, surprisingly, in all pipes Virk's [1] maximum drag reduction asymptote is exceeded. In the studied Reynolds number range the friction behaviour can be divided into different regimes depending on the surface finish of the pipes. For $Re < 30{,}000$ the surface finish has no influence on the friction behaviour. Starting at $Re = 30{,}000$ the friction factors in the rough pipes are increased in comparison to the smooth pipe, and for $Re > 50{,}000$ the type of roughness also influences the friction behaviour. In the "k"-type rough pipe the friction factors then increase with the Reynolds number and approach a nearly constant value at $Re > 75{,}000$. At high Reynolds numbers, therefore, the surface finish has a similar influence on the friction behaviour as it was found for water and the dilute polymer solution. In small angle neutron scattering experiments [40, 41] it was found that a nearly complete alignment of the rod-like micelles in flow direction is responsible for the high drag reduction.

Due to the occurrence of the high drag reduction the question arises whether the

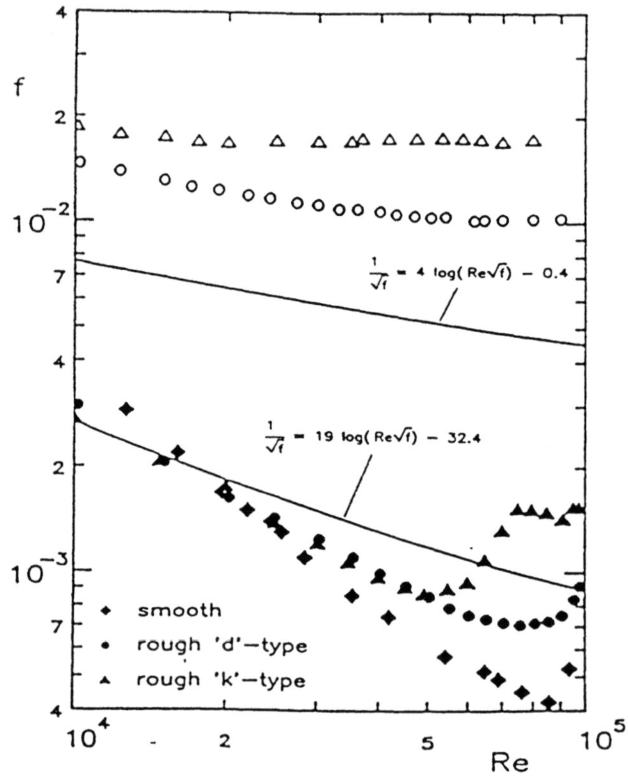

Fig. 15. Friction behaviour of a C_{14}TASal-solution at $c = 945$ ppm and $T = 21°C$; open symbols: water; full symbols: surfactant solution.

flow of these surfactant solutions is turbulent or laminar, because when the Reynolds number would be calculated with a shear viscosity of 1.7 mPa s the data at low Reynolds numbers would fall on Hagen-Poiseuille's law ($f = 16/Re$). In the SIS this value of the shear viscosity is exceeded. The question can be answered by analyzing measured velocity profiles. Therefore in the plots of the velocity profiles of the surfactant solutions the line of a laminar velocity profile has been added. In dimensionless form using U^+ and y^+ the expression for a laminar velocity profile of a pipe flow becomes

$$U^+ = y^+ - \frac{y^{+2}}{(8\,Re)^{1/2}}. \tag{22}$$

In this equation y^+ is based on the solvent viscosity. As shown in Fig. 16 the velocity data of the surfactant solution fall between Virk's ultimate profile, Eq. (18), and a laminar profile, Eq. (22). The measured velocity profiles at $Re = 50{,}000$ elucidate that the different spacing of the roughness elements has no hydrodynamic influence on the

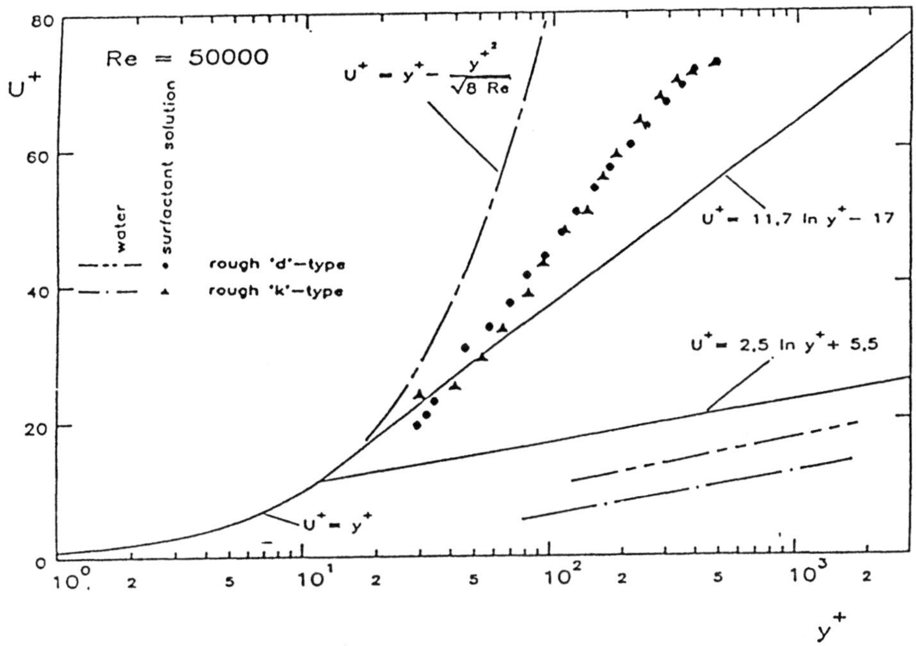

Fig. 16. Dimensionless velocity profiles of a C_{14}TASal-solution at $c = 945$ ppm and $T = 21°C$.

flow of the surfactant solution. The velocity profiles which are identical for both types of roughness element spacings can be approximated by a logarithmic expression, Eq. (13), with a slope of $A = 19.5$. This is an indication of a turbulent flow. Another indication of this is the occurrence of velocity fluctuations in the LDA-signal of the axial velocity component. The velocity data elucidate that the turbulence structure is changed in the whole cross-section of the pipe in comparison to a Newtonian fluid.

In Fig. 17 two velocity profiles obtained in the "k"-type rough pipe are presented.

In the "k"-type rough pipe the friction factor is higher at $Re = 80{,}000$ than that at $Re = 50{,}000$ (see Fig. 15). The slope of both velocity profiles are identical in the whole range of wall distances. Therefore the increase of the friction factor in the "k"-type rough pipe can be related to the hydrodynamic influence of the roughness in the near-wall region.

When in the "k"-type rough pipe the surfactant concentration is lowered, velocity profiles are obtained which can be divided in two regions. The two slopes are identical with that of Virk's [1] ultimate profile ($A = 11.7$) and that of a turbulent core ($A = 2.5$). These two velocity profiles are similar to those of Hendricks et al. [36], see Fig. 11. They make it evident that the hydrodynamic influence of the roughness elements on the near-wall turbulence produces a parallel shift of the velocity profiles to lower dimensionless velocity values.

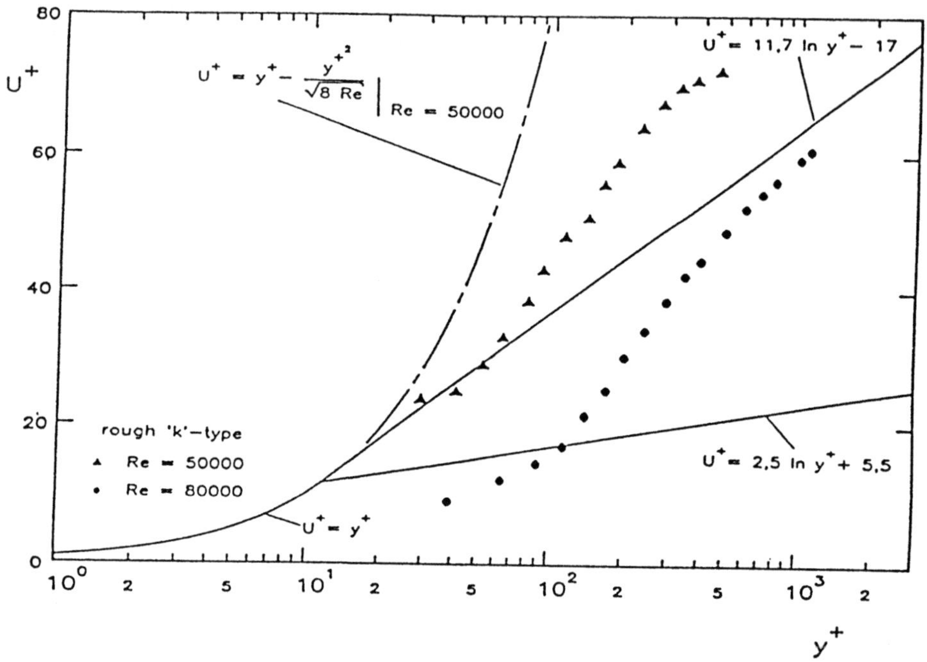

Fig. 17. Dimensionless velocity profiles of a C_{14}TASal-solution at $c = 945$ ppm and $T = 21°C$.

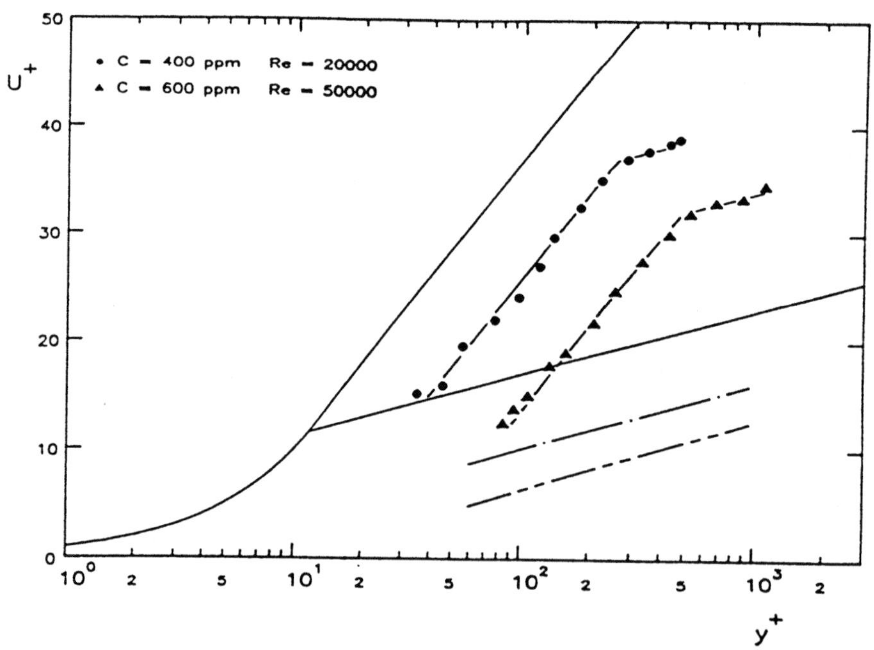

Fig. 18. Dimensionless velocity profile of a C_{14}TASal-solution in a "k"-type rough pipe at $T = 19.5°C$.

4. Conclusions

Experimental data are presented on the friction behaviour of dilute drag reducing polymer and surfactant solutions in a smooth, a "d"-type, and a "k"-type rough pipe. They elucidate that the achievable drag reduction significantly depends on the pipe wall geometry, the Reynolds number, and the additive concentration. For dilute polymer solutions the drag reduction in a smooth pipe is significantly higher than in a rough pipe at the same flow-rate or friction velocity. The onset-point of drag reduction is shifted to higher wall shear stresses with increasing influence of the roughness. In addition, in dilute polymer solutions the maximum attainable drag reduction decreases with increasing influence of the roughness. In the "k"-type rough pipe the friction factor becomes constant at high wall shear stresses designating a fully rough flow regime, a tendency, which already was observed in an earlier study [34].

Since the slopes of the dimensionless velocity profiles were not affected by the surface finish it can be concluded that the changes in the structure of turbulence in drag reducing fluids are not influenced by the roughness for $y^+ > 50$. The surface finish has only an influence on the near-wall region of the flow. Therefore, the influence of the roughness elements is responsible for the negative parallel shift of the dimensionless velocities in the core region and in the elastic sublayer. This is a strong indication that essential features of the turbulent structures in the near-wall region are also determined by the characteristics of the surface finish in drag reducing fluids.

The coherent structures of turbulence close to rough walls exhibit similarities with those occurring close to hydraulically smooth walls for flows of Newtonian fluids [42–45]. Near rough walls a burst-sweep cycle also occurs, but the burst frequency and the momentum transport per burst event are different than for a smooth wall [42, 43].

In the flow of the surfactant solution friction factors were measured which were lower than Virk's [1] maximum drag reduction asymptote. At low Reynolds numbers ($Re < 30{,}000$) there was no influence of the surface finish on drag reduction. This result is in agreement with a result by Virk [14] who found an identical friction behaviour in smooth and rough pipes provided that $k^+ < 12$. In the present experiments this holds for $Re < 30{,}000$. In the region where the drag reduction exceeds Virk's maximum drag reduction asymptote the flow is turbulent. The measured velocity profiles show that in this region a logarithmic layer, similar to Virk's [1] ultimate profile, reaches the centre of the pipe and that the slope of this velocity profile is increased in comparison to Virk's [1] ultimate profile.

Acknowledgement

The authors thank the Deutsche Forschungsgemeinschaft (DFG) for supporting this work by research grant Gi 83/9.

References

1. Virk, P.S., *AIChE J.* 21 (1975) 625.
2. Giesekus, H., Bewersdorff, H.W., Frings, B., Hibberd, M., Kleinecke, K., Kwade, M. and Schröder, R., *Fortschr. Verfahresnstechnik* 23 (1985) 3.
3. Giesekus, H. and Hibberd, M.F., In: Majumdar, A.S. and Mashelkar, R.A., (eds), *Advances in Transport Processes*, Vol. V. New Dehli: Wiley Eastern (1987) p. 229.
4. Shenoy, A.V., *Colloid & Polymer Sci.* 262 (1984) 319.
5. Sellin, R.H.J. and Moses, R.T., *Drag Reduction in Fluid Flows.* Chichester: Ellis Horwood (1989).
6. Gyr, A., *Structure of Turbulence and Drag Reduction.* Berlin: Springer Verlag (1990).
7. Lindgren, E.R. and Hoot, T.G., *Trans. ASME, J. Appl. Mech.* 35 (1968) 417.
8. Sellin, R.H.J., Hoyt, J.W., Pollert, J. and Scrivener, O., *J. Hydraulic Res.* 20 (1982) 235.
9. Mizushina, T. and Usui, H., *Phys. Fluids* 20 (1977) S100.
10. Schümmer, P. and Thielen, W., *Chem Eng. Commun.* 4 (1981) 593.
11. Schmid, A., In: Sellin, R.H.J. and Moses, R.T. (eds), *Drag Reduction.* Bristol: University of Bristol (1984) paper B12.
12. Willmarth, W.W., Wei, T. and Lee, C.O., *Phys. Fluids* 30 (1987) 933.
13. Tiederman, W.G., Luchik, T.S. and Bogard, D.G., *J. Fluid Mech.* 156 (1985) 419.
14. Virk, P.S., *J. Fluid Mech.* 45 (1971) 225.
15. McNally, W.A., Ph.D. Thesis, University of Rhode Island, U.S.A. (1968).
16. Debrule, P. M. and Sabersky, R.H., *Int. J. Heat Mass Transfer* 17 (1974) 529.
17. Spangler, J.G., In: Wells, C.S. (ed.), *Viscous Drag Reduction.* New York: Plenum Press (1969) p. 131.
18. Perry, A.E., Schofield, H.W. and Joubert, P.N., *J. Fluid Mech.* 37 (1969) 383.
19. Nunner, W., *VDI-Forschungsheft* 455 (1956).
20. Möbius, H., *Physikalische Zeitschrift* 41 (1940) 202.
21. Product specifications, Dow Chemical Company.
22. Bewersdorff, H.W. and Ohlendorf, D., *Colloid & Polymer Sci.* 266 (1988) 941.
23. Hopf, L., *ZAMM* 3 (1923) 329.
24. Morris, H.M., *Trans. ASCE* 120 (1955) 373.
25. Nikuradse, J., *VDI-Forschungsheft* 361 (1933).
26. Colebrook, C.F. and White, C.M., *Proc. Royal Soc. London*, Ser. A, 161 (1937) 367.
27. Schlichting, H., *Grenzschicht-Theorie.* Karlsruhe: G. Braun (1965).
28. Hama, F.R., *Trans. Soc. Naval Archit. Marine Engrs.* 62 (1954) 333.
29. Bandhyopadhyay, P.R., *J. Fluid Mech.* 180 (1987) 231.
30. Einstein, H.A. and El-Samni, E.A., *Rev. Mod. Phys.* 21 (1949) 520.
31. Clauser, F.H., In: *Advances in Applied Mechanics*, Vol. 4. New York: Academic Press (1956) p. 1.
32. Moore, W.L., Ph.D. Thesis, University of Iowa, U.S.A. (1951).
33. Perry, A.E. and Joubert, P.N., *J. Fluid Mech.* 17 (1963) 193.
34. Bewersdorff, H.W. and Petersmann, A., *Chem Eng. Commun.* 60 (1987) 130.
35. Ivanyuta, Y.F. and Chekalova, L.A., *J. Eng. Phys.* 31 (1974) 891.
36. Hendricks, E.W., Swearingen, J.D., Horne, M.P. and Lawler, J.V., *AIAA* paper 88-3667 (1988).
37. Ohlendorf, D., Interthal, W. and Hoffmann, H., *Rheol. Acta* 25 (1986) 468.
38. Wunderlich, A.M. and Brunn, P.O., *Colloid & Polymer Sci.* 267 (1989) 289.
39. Vissmann, K. and Bewersdorff, H.W., *J. Non-Newtonian Fluid Mech.* 34 (1990) 289.
40. Lindner, P., Bewersdorff, H.W., Heen, R., Sittart, P., Thiel, H., Langowski, J. and Oberthür, R., *Progr. Colloid & Polymer Sci.* 81 (1990) 107.
41. Bewersdorff, H.W., Dohmann, J., Langowski, J., Lindner, P., Maack, A., Oberthür, R. and Thiel, H., *Physica B* 156 & 157 (1989) 508.
42. Grass, A.J., *J. Fluid Mech* 50 (1971) 233.
43. Sabot, J., Saleh, I. Comte-Bellot, G., *Phys. Fluids* 20 (1977) S150.
44. Ligrani, P.M. Moffat, R.J., *J. Fluid Mech.* 162 (1986) 69.
45. Bandhyopadhyay, P.R. Watson, R.D., *Phys. Fluids* 31 (1988) 1877.

Applied Scientific Research **50**: 369–385, 1993.
K. Krishna Prasad (ed.), Further Developments in Turbulence Management.

Effect of external manipulators on the heat transfer on a flat plate turbulent boundary layer

A. HAMDOUNI & J.P. BONNET
CEAT/LEA URA CNRS 191, 43 rue de l'aérodrome, 86036 Poitiers Cedex, France

1. Introduction

It is now generally agreed that outer layer manipulations are not very efficient for obtaining a definite total drag reduction in turbulent boundary layers: the presence of manipulators creates an amount of drag that practically balances the decrease of wall skin-friction. Hence, the total effect can be considered to be practically nil. This aspect has led to a lack of interest for systematic studies on the process for reducing drag [1].

However, in spite of no proved efficiency in drag reduction [2], let us recall that, provided the form of manipulator is adequately designed, there is no significant bad effect over total drag due to insertion of such devices in the boundary layer.

Substantial alterations of the structure of the layer and appreciable reduction of the skin friction are obtained, and secondary effects such as effects on wall pressure fluctuations or on heat transfer can be of practical interest. As far as the heat transfer is concerned, the manipulation will act as a thermal insulator. Industrial applications can be imagined, like improvement of thermal insulation of bodies placed in hot ambient flows as it can be the case for some rear parts of engines, or, conversely to preserve the temperature of fluid in presence of walls of very different temperatures. This kind of passive thermal insulation can avoid special surface treatment or limit the use of expensive insulating materials.

Indeed, it has been shown that manipulation produces a damping of the turbulent interaction between the internal and external parts of the boundary layer determined by the manipulator wake. The modification of the turbulent dynamic mixing can directly affect the heat transfer. As previously mentioned, this effect can be of interest in engineering. Trigui and Guezennec [4] confirmed this fact; their measurements concern a turbulent boundary layer manipulated in a tandem configuration acting on a wall uniformly heated from the starting end. Their measurements show a reduction of up to 25% of the Nusselt number and also that the relaxation of the heat transfer is faster than that of the dynamic transfer. More recently, Suzuki et al. [5] determined, in a parametric study with just one manipulator, that the maximum reduction of the Nusselt number is of an order of 20%. These results presented by the above mentioned authors do not show any noticeable effect on the mean temperature profiles, at least for the distances from the trailing edge which are not very close to the manipulator. However, it can be supposed that the LEBU can behave like a thermal insulator, thus introducing a plateau in the profile of the mean temperature field. Further experimental studies are required in order to define more precisely the total effects of the manipulators on the development of the thermal boundary layer.

The purpose of the present work is two-fold.

- The first provides experimental data on several aspects of heat transfer modifications due to the presence of a LEBU in turbulent boundary layers. These studies concern the analysis of the wake of a *heated LEBU* in an isothermal environment. These experimental results are complementary to the data already given [7]. Secondly, we give new mean temperature measurements over a *heated wall* in a natural and in a manipulated configuration. These results will be complementary to other experimental results obtained by other authors in comparable configurations [4, 5].
- The second is to test closure models for temperature fluxes in this type of out-of-equilibrium flows. This numerical study will, after validation, be used for parametric studies of the modification of heat transfer over a flat plate turbulent boundary layer in view of other potential industrial applications.

2. Experimental study

2.1. Experimental set-up and flow configurations

Wind tunnel

The experiments were carried out in the subsonic $30 \times 30\,\mathrm{cm}^2$, 26 m/s wind tunnel at the CEAT. The details of the experimental set-up and the measurement procedure are given in [7, 8]. x is the streamwise distance, x_0 corresponds to the location of the manipulator trailing edge and y is the distance normal to the wall. The boundary layer was fully turbulent at the location of the manipulator, the conventional dynamical boundary layer thickness ($\delta_{u_0} = \delta_{.995}$) being 23.5 mm. The local shape factor was 1.35 and the Reynolds number based on the momentum thickness, R_{e_θ} equal to 4000.

Manipulator

All the results presented were obtained with a single steel flat plate manipulator with a rectangular section: chord $c = 25.4$ mm, length $l = 25$ mm, thickness $t = 0.1$ mm.

Table 1 and Fig. 1 give the geometrical characteristics of the different experimental arrangements.

Measurement apparatus

Probes are moved by a stepping motor. The complete data acquisition and traversing mechanism are described in detail in Lemay et al. [8] who studied in depth the isothermal manipulated flow by means of Constant Temperature Hot Wire Anemometry (single and X wires).

The present measurements of the thermal fields were performed by means of thermocouples to obtain the mean values of all the flows. A specially designed dual

Table 1.

	Heated LEBU	Heated wall	
Cases	A	B_1	B_2
U_e(m/s)	26	26	26
δ_{uo} (mm)	23.5	28.	28.
c/δ_{uo}	1.06	0.9	0.9
h/δ_{uo}	0.43	0.36	0.2
$X_c - X_0$ (mm)	/	−278	−278
ξ_c	/	−11.8	−11.8
δ_{to} (mm)	/	13.2	13.2
h/δ_{to}	/	0.76	0.45
$\Delta\Phi_l = T_l - T_e$ (°C)	21	/	/
$\Delta\Phi_w = T_w - T_e$ (°C)	/	16.1	16.1

probe was made with a thermocouple and a pitot tube. The thermocouple was made of chromel-alumel and had diameter of 0.2–0.3 mm at the welding; the pitot tube had an internal diameter of 0.5 mm.

The fluctuations were measured with Constant Current Hot Wire Anemometers in the case of the *heated LEBU*. A special numerical frequency response correction is applied to the signals, in order to adapt the bandwidth.

Thermal configurations

Heated LEBU (*case A*): In the case of the *heated LEBU* the manipulator was slightly overheated at $\Delta\Phi_l = T_l - T_e = 21°$, with T_l the LEBU temperature and T_e the external temperature.

This overheating was obtained by direct electrical heating: a constant voltage was

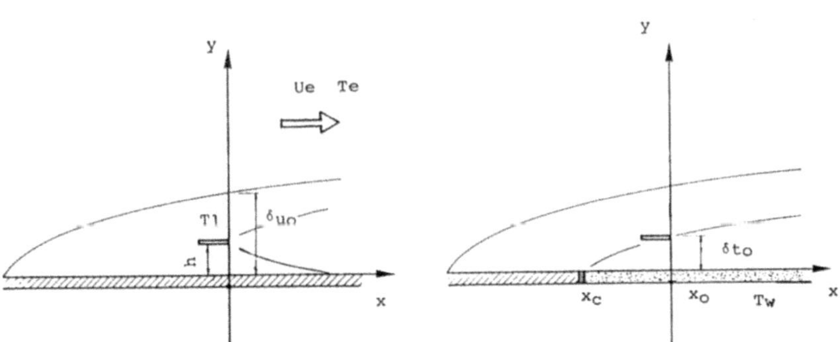

Case A:
Heated LEBU (unheated wall, $T_w = T_e$ & $T_l > T_w$)

Case B:
Heated wall from $x > x_c$

Fig. 1. Schematic description of the flow arrangement.

applied directly to the steel flat plate manipulator. The temperature of the LEBU could then be assumed to be constant.

Heated wall (*case B*): In the case of the *heated wall* below the manipulator, a thermostated water circulation directly inside the wall produced the heat. The water temperature was servo-controlled so that the difference of the temperature between the wall and the incoming air remained constant. The regulation of the temperature of the water was accurate to within less than $\pm 0.1°$. 18 thermocouples located at different location of the streamwise gave information concerning the two-dimensionality and the streamwise uniformity of the wall temperature. The distribution was assumed to be satisfactory, with differences temperature of less than 0.5°C in the streamwise direction of the flow and for at least 85% of the span.

The difference in temperature $\Delta\Phi_w = T_w - T_e$ was equal to 16.1°. In this configuration, only the mean temperature field was measured with the dual Pitot-thermocouple probe.

Several combinations of x_c and h were studied (i.e. starting end of the wall heating, $\xi_c = (x_c - x_0)/\delta_0$, and the height the manipulator).

2.2. Experimental results

For the two experimental configurations, the temperature difference being quite small, the thermal field could be considered as a passive marker. As an example, in the case of the heated LEBU, the Richardson number based on the length of the manipulator was of the order of 4.10^{-5}; this small value allowed free convection effects to be neglected.

2.2.1. Case A: heated LEBU

The manipulator was placed at $h/\delta_{u0} = 0.43$ from the wall. This location corresponded to the detailed measurements of Lemay et al. [8].

The profiles of the mean temperature field are presented in terms of the non-dimensional variables y/δ_{u0} (normal to the wall direction) and $\xi = (x - x_0)/\delta_{u0}$ in the streamwise direction. The main objective of the present work being to test closure models for the thermal field, we are not discussing in detail here the mechanisms of the turbulence structure modifications that can be deduced from these measurements. A more complete discussion of these results will be found in [7].

2.2.2. Case B: heated wall

No effect on the mean temperature profile was observed when the wall heating started at location $\xi_c = 0$. Indeed, in this case the manipulator wake was very weak when it reached the thermal boundary layer which started from $\xi_c = 0$. We present here only two cases (noted B1 and B2) where the heating of the wall started upstream from the location of the manipulator. For these two cases, depending on the distance of the manipulator from the wall, the thermal boundary layer either just reached the LEBU (B1) or the manipulator was close enough to the wall to be entirely enclosed in the thermal layer (B2). In the two cases, the dynamical wake of the LEBU interacted with the thermal layer. In other words, in case B1, h was of the order of δ_{t_0} (thermal boundary layer thicknesses at the manipulator location) whereas in case B2, h was about half δ_{t_0} (see Table 1).

Figs 2a and 2b show typical results of the two configurations ($h/\delta_{u_0} \simeq 0.36$ and 0.2, h/δ_{t_0} being respectively $\simeq 0.76$ and 0.45). A slight "plateau effect" in the near wake of the manipulator ($\xi \simeq 0.11 - 1.12$) can be observed. The manipulator wake acts like a *thermal insulator* between the external and the internal parts of the boundary layer, but this effect is limited to the immediate vicinity of the LEBU. It can be supposed that this effect has a strong influence on the heat transfer coefficient H. This effect has not been quantitatively analyzed up to now for the present experiment because the measurement technique is not suited for this purpose. However, heat transfer decreases have already been cited by other authors ([4, 5]) by direct measurements of H.

The significance and the extent of the plateau depends very much on both h/δ_{u_0} and h/δ_{t_0}. Mainly in the case B2, the plateau on the temperature profile is quite clear (see Fig. 2b) but rapidly disappears; this is no longer significant for distances greater than a few initial boundary layer thicknesses. This particular insulation effect has not been

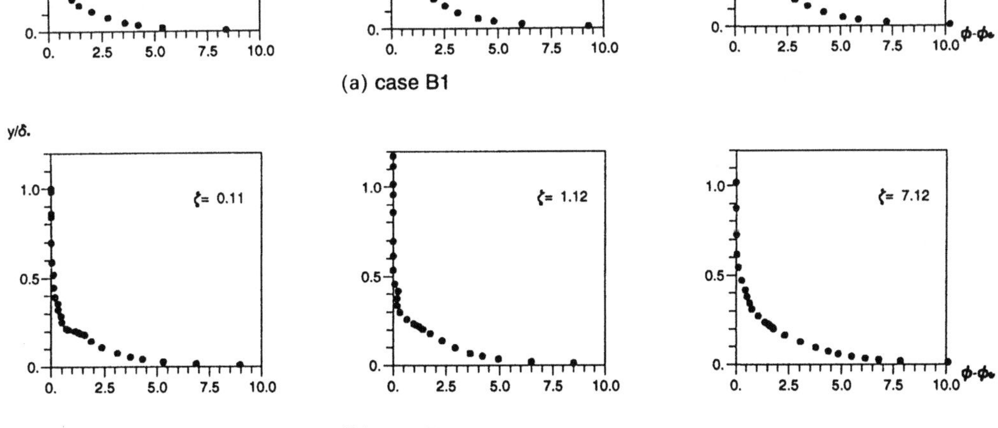

Fig. 2. Mean temperature measurements on the heated wall cases.

shown yet by the experiments of the same type carried out by Trigui and Guezennec [4]; and Suzuki et al. [5]. These authors do not present mean temperature profiles in the immediate vicinity of the manipulator.

3. Calculations

3.1. Introduction

The numerical prediction of the flow in LEBU manipulated turbulent boundary layers has been performed by numerous research groups using different approaches. We quote [2, 6] for recent reviews of these calculations. Our purpose being to analyze the thermal fields in heated configurations, no attempts have been made to develop a new code for the prediction of the dynamic field. The present calculations of the velocity field are based essentially on the code built, validated and used currently at ONERA-CERT by Tenaud ([9, 12]). This author tested several closure models for the turbulence transport equations, from $k - \varepsilon$ up to full Reynolds Stress Model. We retain here the conventional closure model like for instance the three transport equation model $(k, \varepsilon, \overline{uv})$ [9, 12].

As has been already stated, the temperature can be considered as a passive field, whose development can be calculated *a-posteriori*, if the velocity field for a given configuration is known. Taking into account the small excess of temperature of the manipulator or of the wall, this hypothesis can be admitted provided that the temperature level ($\leqslant 320\,K$) is low enough to avoid noticeable influence on the mean properties of the fluid (density or viscosity). On the other hand, the Richardson number is small, thus the effects of the free convection compared to those of the forced convection can be neglected. Consequently, it can be considered that the thermal field does not modify the dynamic field.

The calculations are then performed in two steps: the dynamic fields are calculated using a semi-elliptic code. In a second step, the thermal field is calculated for several levels of closure.

3.2. Prediction models for the passive scalar

As is the case for the dynamic field, the assumption of the conventional stationarity and of the two-dimensionality are used for the thermal field. In addition, as is usual in these problems, the streamwise diffusion can be neglected compared to the transverse diffusion. The mean temperature equation is then

$$U \frac{\partial \Phi}{\partial x} + V \frac{\partial \Phi}{\partial y} = \chi \frac{\partial^2 \Phi}{\partial y^2} - \frac{\partial}{\partial y} (\overline{v\phi}). \tag{1}$$

Two closure models will be presented. One will use a first order closure, whereas the other will make use of a single transport equation for $\overline{v\phi}$.

3.2.1. *Model 1: First order model*

This quite simple model is based on the Reynolds analogy, leading to the introduction of a turbulent Prandtl number $\mathscr{P}_t$. This closure is then

$$\overline{v\phi} = -\chi_t \frac{\partial \Phi}{\partial y}. \tag{2}$$

And

$$\chi_t = c_\mu \frac{k^2}{\mathscr{P}_t \varepsilon} f(R_t)$$

the constant c_μ has the normal value 0.09.

As usual, a wall function $f(R_t)$ has been introduced in order to take into account the low values of the Reynolds number obtained close to the wall (close to the LEBU or the wall, typically when $y^+ \leqslant 60$):

$$f(R_t) = \exp\left(\frac{-3.4}{\left(1 + \frac{R_t}{50}\right)^2}\right),$$

with $R_t = (\rho k^2/\mu\varepsilon)$.

In order to maintain a substantial simplicity of this closure we use a constant turbulent Prandtl number, with the conventional value of $\mathscr{P}_t = 0.89$.

3.2.2. *Model 2: Second order closure*

Model 2 is a model which uses a transport equation for $\overline{v\phi}$ in the external zone of the turbulent boundary layer defined by $y^+ > 60$. This equation is linked to the first order model, already described, in the internal zone of the boundary layer defined by $y^+ \leqslant 60$. The transport equation of $\overline{v\phi}$ can be obtained thanks to the classic model described by Launder [11]. If the diffusion along the main direction of the flow relative to the diffusion along the normal direction of the wall is neglected, and if it is supposed moreover that $(\partial V/\partial x)$ can be neglected then the transport equation of the $\overline{v\phi}$ can be written

$$U \frac{\partial \overline{v\phi}}{\partial x} + V \frac{\partial \overline{v\phi}}{\partial y} + A\overline{v\phi} - \frac{\partial}{\partial y}\left(D_t \frac{\partial \overline{v\phi}}{\partial y}\right) = P_\phi, \tag{3}$$

with

$$A = \left(c_{\phi 1} \frac{\varepsilon}{k} - (1 - c_{\phi 2}) \frac{\partial U}{\partial x} \right),$$

$$D_t = \frac{c_\mu}{\sigma_\phi} \frac{k^2}{\varepsilon},$$

$$P_\phi = - \left(\overline{uv} \frac{\partial \Phi}{\partial x} + \overline{v^2} \frac{\partial \Phi}{\partial y} \right).$$

The constants of the model are those normally used [11], $c_{\phi 1} = 2.9$, $c_{\phi 2} = 0.4$ et $\sigma_\phi = 0.3$.

3.3. Numerical procedure

3.3.1. Summary of the procedure used for the dynamic problem

The dynamic problem was solved with the code developed by Tenaud and described in detail [9, 12]. The equations having a conservative form, the finite volume method of Patankar et al. [15] was employed. Calculations ranged from $\xi = -15$ to 40. The transverse domain reached $20\delta_{u0}$ from the wall. Close to the wall and to the manipulator, the mesh was refined, the resulting grid size was 110×160 and the LEBU lay on 10×4 mesh points. The semi-elliptic system was solved with an X-marching procedure. The pressure is specified at the last location and the iterative process ran until the pressure field converged. As shown by Tenaud, the initial conditions was not crucial, since they were applied sufficiently upstream from the LEBU location to be sure that the boundary layer was in a state of equilibrium at the location of the manipulator.

3.3.2. Scalar field prediction with the first order model

The equation for the mean field and the associated boundary conditions are

$$\begin{cases} \operatorname{div}(\underline{U}\Phi) - \dfrac{\partial}{\partial y} \left\{ (\chi + \chi_t) \dfrac{\partial \Phi}{\partial y} \right\} = 0 \quad \text{in } \Omega, \\ \Phi|_{\Gamma_i} = \Phi_i, \qquad \Phi|_{\Gamma_e} = \Phi_e, \\ \alpha\Phi + \beta \dfrac{\partial \Phi}{\partial n}|_{\Gamma_w \cup \partial\Delta} = \delta. \end{cases} \tag{4}$$

The domain Ω is defined by: $\Omega =]0, X_M[\times]0, Y_M[\backslash \Delta$.

The manipulator represents the domain Δ. The different domains and their boundaries are defined in Diagram 1. α, β and δ are functions in $L^2(\Omega)$ such that

$$\begin{cases} \beta(M) = 1 - \alpha(M), \\ \delta(M) = \alpha(M).\Phi_r. \end{cases}$$

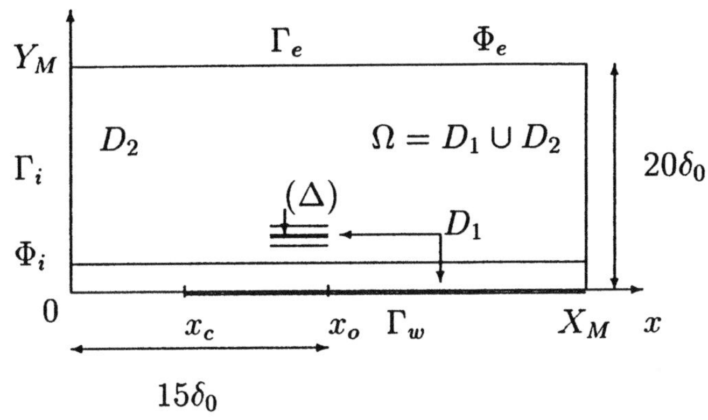

Diagram 1. Definitions of the numerical integration domain.

The trace of α on $\partial\Omega$ allows the different boundary conditions to be defined. Table 2 describes the corresponding conditions.

Case A corresponds to the adiabatic wall, i.e. $(\partial\Phi/\partial y)_{\Gamma_w} = 0$, the manipulator being overheated with a temperature drift of $\Delta\Phi_l$.

Case B deals with a non-heated manipulator. The thickness of the LEBU being quite small, we assume that the manipulator is always in thermal equilibrium with the ambient air, i.e. $(\partial\Phi/\partial y)_{\partial\Delta} = 0$. The wall is heated at a constant temperature at a distance $x_c(\xi_c)$ from the manipulator. The initial condition at ξ_i is then a constant value $\Phi_0 = \Phi_e$ if $\xi_c > \xi_i$. If the heating of the wall occurs upstream of the beginning of the computation domain ($\xi_c < \xi_i$), the initial condition is given by experiments or by theoretical considerations. In the present calculations, when $\xi_c < \xi_i$, we consider the similarity profile as the initial condition for the mean temperature profile.

3.3.3. Scalar field prediction with the second order model

The equation for the mean scalar is solved in the domain Ω. The boundary and initial conditions are the same as for the first order model.

The turbulent equations are solved in the domain D_2, for which $y^+ > 60$. Close to the walls, corresponding to D_2, the eddy diffusivity model previously described is used. Continuity of $\overline{v\phi}$ is imposed at the interface between D_1 and D_2. On Γ_e and on $\partial\Delta$, we assume that $\overline{v\phi} = 0$. The initial conditions for $\overline{v\phi}$ are obtained from the eddy diffusivity model.

Table 2. Definition of boundary conditions.

Case	$\alpha\|_{\Gamma_w}$	$\alpha\|_{\partial\Delta}$	Φ_r
A (Heated Lebu)	0	1	$\Phi_e + \Delta\Phi_l$
B (Heated Wall)	$H(x - x_c)$	0	$\Phi_e + \Delta\Phi_w$

$H(x - x_c)$ is the Heaviside function.

3.3.4. Numerical procedure

Formally, the two above mentioned models can be written as a single set of equations:

$$\left\{\begin{array}{ll} \operatorname{div}(\underline{U}\Phi^{(n)}) - \dfrac{\partial}{\partial y}\left\{(\chi+\chi_t)\dfrac{\partial\Phi^{(n)}}{\partial y}\right\} = S(\Phi^{(n-1)}, \overline{v\phi}^{(n-1)}) & \text{in } \Omega, \\ \operatorname{div}(U\overline{v\phi}^{(n)}) - \dfrac{\partial}{\partial y}\left(D_t\dfrac{\partial\overline{v\phi}^{(n)}}{\partial y}\right) + A\overline{v\phi}^{(n)} = P_\phi^{(n)} & \text{in } D_2, \\ \overline{v\phi}^{(n)} = F(\phi^{(n)}) & \text{in } D_1, \end{array}\right. \tag{5}$$

where n is the iteration number. The convergence criterion for the mean temperature field is set at a relative level of 10^{-5}.

The third equation represents the wall conditions previously described. Indeed, for the first order model $D_1 = \Omega$ and $D_2 = \emptyset$.

The term

$$S(\Phi^{(n-1)}, Z_\phi^{(n-1)}) = -\left(\frac{\partial\overline{v\phi}^{(n-1)}}{\partial y} + \frac{\partial}{\partial y}\left\{(\chi+\chi_t)\frac{\partial\Phi^{(n-1)}}{\partial y}\right\}\right)$$

is dropped for the first model, the scheme being entirely explicit. When model 2 is considered, the term

$$\chi_t\frac{\partial\Phi^{(n)}}{\partial y}$$

is, as usual, added to improve the stability of the scheme.

The conservative form of these equations allows the finite volume approach to be used. The transverse convective terms are discretized following a hybrid scheme. The equations for the scalar field being parabolic, an X-marching method is used.

3.4. Results

First, the calculations are checked for validation on a natural, non-manipulated boundary layer. The predictions of the heat transfer coefficient and the analogy factor are in quite good agreement with the empirical $-1/5$ power law.

3.4.1. Predictions in the experimental test case A: heated LEBU

Figure 3 shows the comparisons between the calculations and the experiments for models 1 and 2. The comparisons are presented at $7.84\delta_{u0}$ downstream from the trailing edge. It can be noticed that the first order model predicts quite satisfactorily the value of the maximum of the temperature excess in the wake of the manipulator. However, it can be observed that in the external part of the boundary layer the diffusion of the temperature is underestimated by the model. This situation is

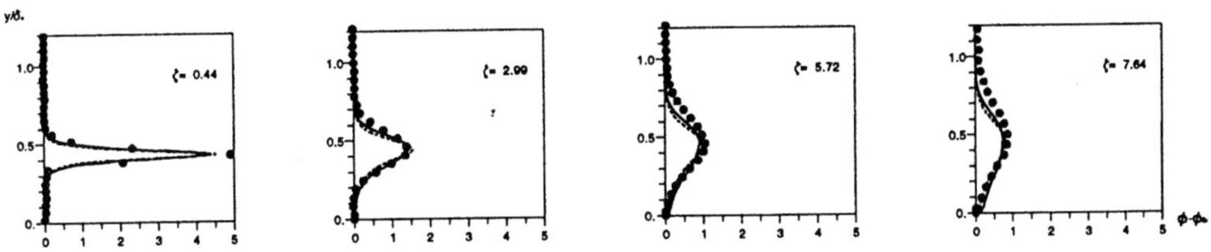

Fig. 3. Comparisons between predictions and experiments. Case A. (● experiments, - - - - - - - Model 1, —— Model 2).

somewhat bettered by the second order model, as presented in Fig. 3. However whatever the model, the predictions underestimate the growth of the thermal wake, particularly in the external part of the manipulated boundary layer. Nevertheless, even the simplest model is able to predict correctly the main features of this manipulated flow, such as the maximum temperature and where the wake reaches the wall. This location has been proved to be quite important in the determination of drag reduction processes; indeed, it corresponds to the location of the maximum reduction of skin friction coefficient.

If we consider the performance of the model for predictions of the dynamic characteristics of this flow, we can observe, according to Tenaud et al. [12], that the model overpredicts the turbulent diffusion $\overline{uv}$ in the external part. Fig. 4, taken from [12], shows that this overprediction can be quite significant (it should be reminded that the overall characteristics of the flow are satisfactorily predicted, like, for example, the skin friction coefficient).

The errors in the prediction of both the dynamic and thermal fields occur in the external part of the layer, but with contrary tendencies. More precisely, we can notice that the thermal diffusion is underestimated (see Fig. 3). This effect is in contradiction with the dynamic behaviour since the overprediction of $\overline{uv}$ results in an overprediction

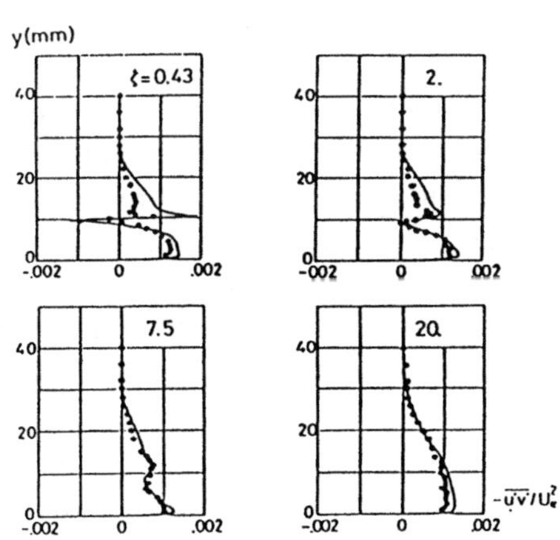

Fig. 4. Comparisons between experiments (●) and the three equations model prediction after Tenaud [12].

of the dynamic diffusion. These different behaviours suggest that two distinct scales should be used – one for the scalar field and the other for the dynamic field. Such an approach is currently developed by Craft and Launder [13, 14]. The use of a variable ratio between time scales for temperature and velocity fluctuations could improve the present predictions.

3.4.2. Predictions in the experimental test case B: heated wall, present experimental case

In this application, the wall was heated immediately downstream from the beginning of the computation domain: $x_i < x_c$. Then, the thermal boundary layer developed *inside* the dynamic turbulent boundary layer. The thermal thickness was then lower than the dynamic thickness.

Computations in the absence of a manipulator (natural case) were performed in order to validate the predictions. Fig. 5a presents the mean temperature profiles obtained with the first order model and the comparison with the experimental results. It can be observed that the development of the thermal layer is in good agreement with the experiments. The predictions of the second order model are quite comparable with the first order model. For comparison, the mean velocity profile predicted at a location close to the LEBU trailing edge ($\xi = 0.21$) is compared with the experiments. It can be observed that at the location of the manipulator, the thermal layer occupies approximately half the transversal value of the dynamic boundary layer thickness. The thermal conditions computed immediately upstream from the LEBU are then quite representative the experimental conditions.

With the manipulator, we present here only the comparisons in the case B_2, i.e. when the manipulator is entirely embedded in the thermal layer: $h/\delta_{t_0} = 0.45$ and $h/\delta_{u_0} = 0.2$. This case corresponds to the more obvious thermally blocking effect of the manipulator already described. Fig. 6 presents the comparisons between the experiments and the calculations with the two models. Apparently the first order model predicts quite well the plateau effect experimentally observed. The relaxation occurring downstream is also well reproduced by the model. Although more elaborate, the second order closure model overestimates the plateau effect, and the thermal field for the locations further away are not as well predicted as in the case of the first model. This result is quite unexpected and tends to show that the turbulent temperature-velocity correlation $\overline{v\phi}$ is probably more difficult to predict close to the manipulator. Owing to the lack of experimental results for this quantity, we are not at present able to discuss the behavior of the second order model as far as the turbulence is concerned. Contrary to the requirement of the dynamic prediction, a simple model appears to be sufficient to predict correctly the thermal field.

3.4.3. Predictions in the experimental test case B: heated wall, Suzuki et al. case

We have also applied the prediction code to the configuration experimentally studied by Suzuki et al. [5]. The wall temperature was 20° C higher than the external one, the wall being heated from its starting point. The heat transfer coefficient was obtained by

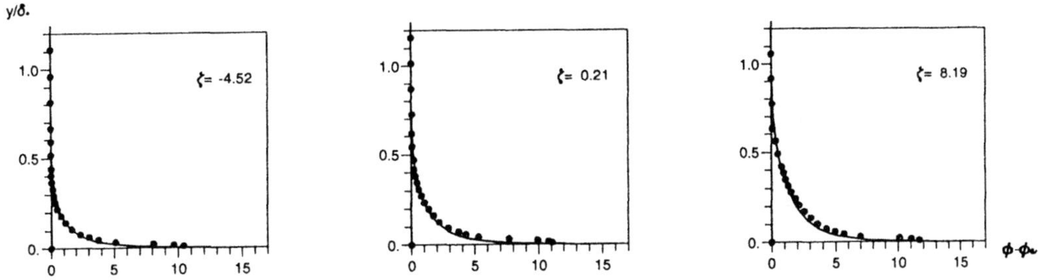

a) Thermal field

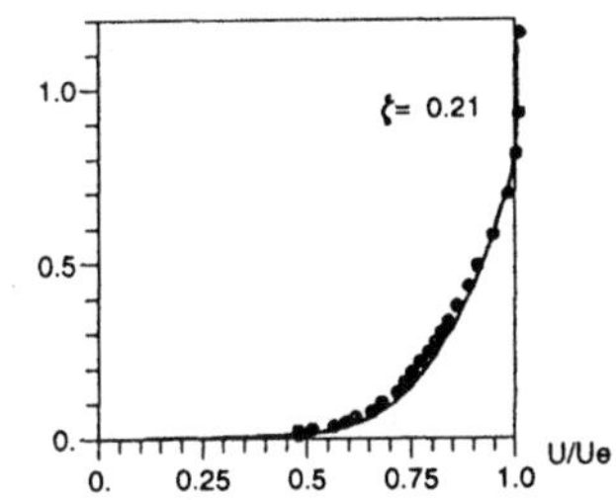

b) Mean velocity field

Fig. 5. Test of Model 1 in unmanipulated case (● experiments, —— prediction).

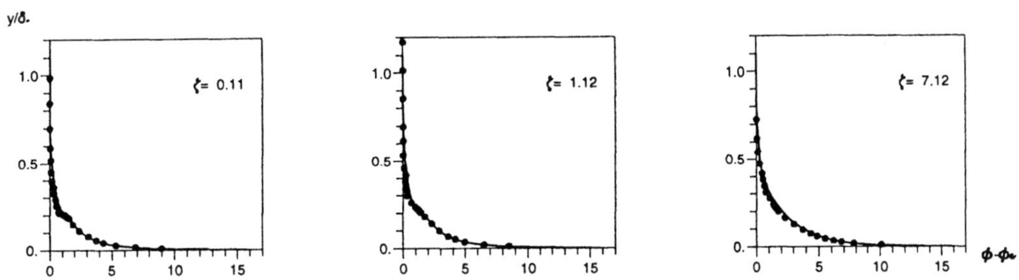

a) Model 1

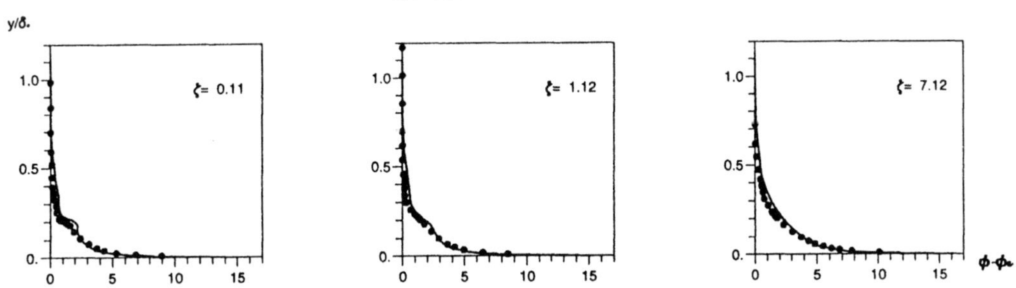

a) Model 2

Fig. 6. Mean temperature profil for the heated wall case (● experiments, —— prediction).

direct estimation of the heat equilibrium of the electrical energy used for heating the wall. The main results concerning the heat transfer coefficient H obtained by Suzuki et al. can be summarized as follows:

– For a given l/δ_{uo}, the maximum of reduction in heat transfer was observed for a LEBU distance of $h/\delta_{uo} = 0.25$ from the wall. The coefficient H evolves towards the value of an undisturbed case more closer to the LEBU when the manipulator is closer to the wall. For example, the relaxation is achieved at $\xi \simeq 15$ when $h/\delta_{uo} = 0.25$ and at $\xi \simeq 5$ when $h/\delta_{uo} = 0.07$.

– For a given h/δ_{uo}, the reduction of the heat transfer coefficient was better when l/δ_{uo} increased.

In every case, it is clear that the minimum of the heat transfer coefficient occurs at shorter distances from the LEBU than the minimum of skin friction coefficient does. Therefore the relaxation of the thermal boundary layer seems to be faster than that of the dynamic layer.

Figure 7 presents some of the results used for the comparisons with the present calculations (manipulator chord $c = \delta_{uo}$, $h = 0.25\delta_{uo}$). The results presented in Fig. 7a concern the skin friction reduction. The usual admitted maximum of reduction of skin friction can be observed. However, it should be emphasized that this maximum occurs at shorter downstream distances from the manipulator than have been generally reported by several authors. As an example, in similar conditions, Coustols et al. [10] report that maxima of skin friction reduction occur at $\xi \simeq 12$; a second particular behavior of the experiments of Suzuki et al. is that the relaxation of C_f is more rapid than what is generally observed. These discrepancies with other authors are not yet clearly understood and require a careful analysis of the experimental conditions.

Figure 7a gives the results of the calculations for the skin friction coefficient. The values of C_f are normalized by their natural values at the same location (C_{f_0}). The

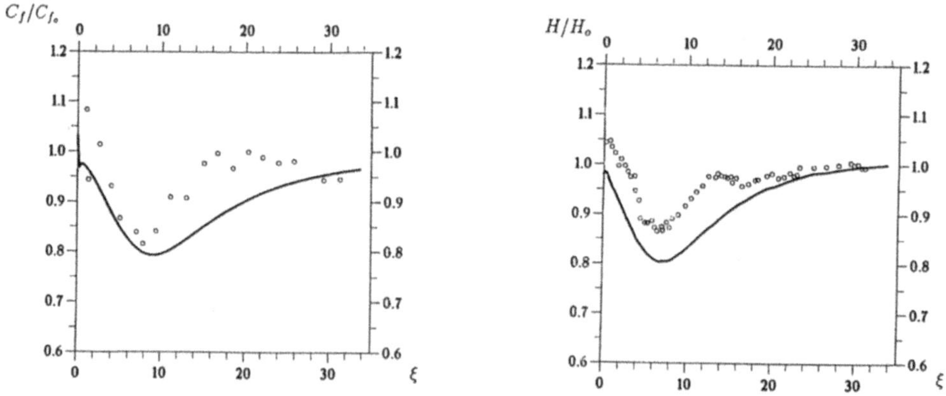

a) Skin friction coefficient. b) Heat transfer coefficient.

Fig. 7. Prediction of the skin friction and heat transfer for the heated wall case of Suzuki [5]. (○ Suzuki's experiments, —— Model 1 prediction).

calculations reproduce the maximum of skin friction reduction with a slight overestimation of the predictions ($C_f/C_{f_0} \simeq 0.8$ instead of 0.83). However, the relaxation in the calculations is slower than with the experiments. The results of the computations are nevertheless in better agreement with the rate of relaxation generally admitted, which is, as previously recalled, slower than with the experiments of Suzuki et al.

The heat transfer function,

$$H = -\frac{\chi}{(\Phi_w - \Phi_e)U_e}\left(\frac{\partial \Phi}{\partial y}\right)_w,$$

is calculated with the first order model tested with some success on the previous case B_2. As far as the thermal field is concerned, Fig. 7b shows the comparisons of the heat transfer coefficient, also normalized here with its natural value. The heat transfer predictions behave in quite a similar manner to that of the skin friction: the maximum of the reduction is of the order of 0.8, while the experimental values are roughly 0.86.

The comparisons that can be made between dynamic and thermal wall characteristics are quite comparable for the experiments and for the predictions. The maxima are comparable in amplitude, with a systematic overprediction in the computations. A second important aspect is the location where these maxima occur: the heat transfer extremum occurs, as has been proved by the experiments, closer to the trailing edge than for the skin friction. This experimental result is very well reproduced by the present calculations and is compatible with the behavior deduced from the use of the analogy factor obtained with $\mathscr{P} \neq 1$.

4. Conclusion

The present experimental results provide complementary information on the influence of external manipulators on the temperature considered as a passive contaminant in turbulent boundary layers. In the case of a heated LEBU, previous results are cited. Fresh results concern the change downstream of the thermal boundary layer developing inside a turbulent boundary layer in the presence of a manipulator. It has been shown that the LEBU behaves more or less like a thermal insulator. The well known dynamic decorrelation created between the external and the internal parts of the boundary layer performed by the blade has a direct effect on the development of the mean temperature field. A plateau is observed in the temperature field in the immediate vicinity of the manipulator. This effect is in agreement with the wall heat transfer measurements carried out by other authors.

The computations of the flows use the dynamic prediction previously developed by Tenaud. The thermal part of the code allows to study the different configurations: heated LEBU or heated wall. Two models are developed: a first order and a second order closure models.

In the case of the heated LEBU, a constant Prandtl number model underpredicts the diffusion in the external part of the boundary layer; the second order model is in slightly better agreement with the experiments. These defects of the predictions could be compared to certain defects previously observed for the dynamic turbulence predicted in the external part of the boundary layer. The two models are however quite able to predict the excess of the mean temperature in the wake of the manipulator.

In the case of a heated wall, the two models are validated in the natural flow. The effects of an external manipulator are predicted better by the first order model. In this flow configuration, the test of the second order model requires probably more detailed experimental results to validate the turbulent temperature fields. Lastly, the application of the prediction code to the experiments of Suzuki et al. shows that a simple model for the thermal field can be quite sufficient to perform acceptable estimations of the overall effects.

The present experiments have shown some typical manipulator effects on the thermal field that confirm other experimental studies. The associated calculations show that even simple models for the thermal field, used with a three-equation model for the dynamic field, are able to predict the main effects. This conclusion allows to undertake parametric studies on the effects of external manipulators on various thermal configurations.

Acknowledgements

The authors thank M.D. Alem for his contribution to the experimental work and Dr C. Tenaud for providing the dynamic simulation code and for fruitful discussions.

References

1. Choi, K.S., Report on the 5th European drag reduction working meeting. *ERCOFTAC bulletin* (March 1991).
2. Anders, J.B., Outre-layer manipulators for turbulent drag reduction. Viscous drag reduction in boundary layers. *Progress in astronautics and aeronautics*, 123 (1990).
3. Lemay, J.L., Etude experimentale du comportement de la turbulence dans une couche limite incompressible en presence d'un manipulateur externe. Ph.D., Laval (Canada) (1989).
4. Trigui, N. and Guezennec, Y.G., Heat transfert reduction in manipulated turbulent boundary layers. *Int. J. Heat and Fluid Flow* 11 (1990) 214–219.
5. Suzuki, K., Suzuki, H., Inaoka, K. and Kigawa, H., Heat transfert in turbulent boundary layer, with an insertion of a LEBU plate. Munich: T.S.F. (Oct. 1991) pp. 24.5.1–24.5.6.
6. Savill, A.M. and Coustols, E., Turbulent skin friction drag reduction by active and passif means. AGRAD FDP/VKI Special Course on skin friction drag reduction, Belgium (March 1992).
7. Bonnet, J.P., Delville, J. and Lemay, J., Study of LEBUS modified turbulent boundary layer by use of passive temperature contamination. *Proc. R.A.S. Int. Conf. Turbulent Drag Reduction by Passive Means*. London (1987) pp. 45–68.
8. Lemay, J., Bonnet, J.P. and Delville, J., Turbulent kinetic energy balance in a LEBU modified turbulent boundary layer. *Exp. in Fluid* 9 (1990) 301–308.

9. Tenaud, C., Simulation numerique de l'ecoulement autour d'un manipulateur externe de couche limite. Thése, Toulouse, France (1988).
10. Coustols, E. and Cousteix, J., Reduction du frottement turbulent: Modérateurs de la turbulence. *La Recherche Aérospatiale* 2 (1986) 145–160.
11. Launder, B.E., Second-moment closure: present ... and future? *Int J. Heat and Fluid Flow* 10 (1989) 282–300.
12. Tenaud, C., Lemay, J., Bonnet, J.P. and Delville, J., Balance of the turbulent kinetic energy downstream a single flate plate manipulator comparisons between detail experiments and modelling. In: Coustols, E. (ed.), *Turbulence Control by Passive Means.* Dordrecht: Kluwer Academic Publishers (1990).
13. Craft, T.J. and Launder, B.E., A new model for the pressure/scalar-gradient correlation and its application to homogeneous and inhomogeneous free shear flow. T.S.F., Stanford University (1989) pp. 17.1.1–17.1.5.
14. Craft, T.J. and Launder, B.E., Computation of impinging flow using second-moment closures. T.S.F., Munich, Germany (1991) pp. 8.5.1–8.5.6.
15. Patankar, S.V. and Spalding, D.B., Heat and mass transfert in boundary layers, a general calculation procedure. London: Intertext (1970).

Mechanics

FLUID MECHANICS AND ITS APPLICATIONS

Series Editor: R. Moreau

Aims and Scope of the Series

The purpose of this series is to focus on subjects in which fluid mechanics plays a fundamental role. As well as the more traditional applications of aeronautics, hydraulics, heat and mass transfer etc., books will be published dealing with topics which are currently in a state of rapid development, such as turbulence, suspensions and multiphase fluids, super and hypersonic flows and numerical modelling techniques. It is a widely held view that it is the interdisciplinary subjects that will receive intense scientific attention, bringing them to the forefront of technological advancement. Fluids have the ability to transport matter and its properties as well as transmit force, therefore fluid mechanics is a subject that is particularly open to cross fertilisation with other sciences and disciplines of engineering. The subject of fluid mechanics will be highly relevant in domains such as chemical, metallurgical, biological and ecological engineering. This series is particularly open to such new multidisciplinary domains.

1. M. Lesieur: *Turbulence in Fluids.* 2nd rev. ed., 1990 ISBN 0-7923-0645-7
2. O. Métais and M. Lesieur (eds.): *Turbulence and Coherent Structures.* 1991 ISBN 0-7923-0646-5
3. R. Moreau: *Magnetohydrodynamics.* 1990 ISBN 0-7923-0937-5
4. E. Coustols (ed.): *Turbulence Control by Passive Means.* 1990 ISBN 0-7923-1020-9
5. A.A. Borissov (ed.): *Dynamic Structure of Detonation in Gaseous and Dispersed Media.* 1991 ISBN 0-7923-1340-2
6. K.-S. Choi (ed.): *Recent Developments in Turbulence Management.* 1991 ISBN 0-7923-1477-8
7. E.P. Evans and B. Coulbeck (eds.): *Pipeline Systems.* 1992 ISBN 0-7923-1668-1
8. B. Nau (ed.): *Fluid Sealing.* 1992 ISBN 0-7923-1669-X
9. T.K.S. Murthy (ed.): *Computational Methods in Hypersonic Aerodynamics.* 1992 ISBN 0-7923-1673-8
10. R. King (ed.): *Fluid Mechanics of Mixing.* Modelling, Operations and Experimental Techniques. 1992 ISBN 0-7923-1720-3
11. Z. Han and X. Yin: *Shock Dynamics.* 1993 ISBN 0-7923-1746-7
12. L. Svarovsky and M.T. Thew (eds.): *Hydroclones.* Analysis and Applications. 1992 ISBN 0-7923-1876-5
13. A. Lichtarowicz (ed.): *Jet Cutting Technology.* 1992 ISBN 0-7923-1979-6
14. F.T.M. Nieuwstadt (ed.): *Flow Visualization and Image Analysis.* 1993 ISBN 0-7923-1994-X
15. A.J. Saul (ed.): *Floods and Flood Management.* 1992 ISBN 0-7923-2078-6
16. D.E. Ashpis, T.B. Gatski and R. Hirsh (eds.): *Instabilities and Turbulence in Engineering Flows.* 1993 ISBN 0-7923-2161-8
17. R.S. Azad: *The Atmospheric Boundary Layer for Engineers.* 1993 ISBN 0-7923-2187-1
18. F.T.M. Nieuwstadt (ed.): *Advances in Turbulence IV.* 1993 ISBN 0-7923-2282-7

Kluwer Academic Publishers – Dordrecht / Boston / London

Mechanics

SOLID MECHANICS AND ITS APPLICATIONS

Series Editor: G.M.L. Gladwell

Aims and Scope of the Series

The fundamental questions arising in mechanics are: *Why?, How?,* and *How much?* The aim of this series is to provide lucid accounts written by authoritative researchers giving vision and insight in answering these questions on the subject of mechanics as it relates to solids. The scope of the series covers the entire spectrum of solid mechanics. Thus it includes the foundation of mechanics; variational formulations; computational mechanics; statics, kinematics and dynamics of rigid and elastic bodies; vibrations of solids and structures; dynamical systems and chaos; the theories of elasticity, plasticity and viscoelasticity; composite materials; rods, beams, shells and membranes; structural control and stability; soils, rocks and geomechanics; fracture; tribology; experimental mechanics; biomechanics and machine design.

1. R.T. Haftka, Z. Gürdal and M.P. Kamat: *Elements of Structural Optimization.* 2nd rev.ed., 1990 ISBN 0-7923-0608-2
2. J.J. Kalker: *Three-Dimensional Elastic Bodies in Rolling Contact.* 1990 ISBN 0-7923-0712-7
3. P. Karasudhi: *Foundations of Solid Mechanics.* 1991 ISBN 0-7923-0772-0
4. N. Kikuchi: *Computational Methods in Contact Mechanics.* (forthcoming) ISBN 0-7923-0773-9
5. *Not published.*
6. J.F. Doyle: *Static and Dynamic Analysis of Structures.* With an Emphasis on Mechanics and Computer Matrix Methods. 1991 ISBN 0-7923-1124-8; Pb 0-7923-1208-2
7. O.O. Ochoa and J.N. Reddy: *Finite Element Analysis of Composite Laminates.* ISBN 0-7923-1125-6
8. M.H. Aliabadi and D.P. Rooke: *Numerical Fracture Mechanics.* ISBN 0-7923-1175-2
9. J. Angeles and C.S. López-Cajún: *Optimization of Cam Mechanisms.* 1991 ISBN 0-7923-1355-0
10. D.E. Grierson, A. Franchi and P. Riva: *Progress in Structural Engineering.* 1991 ISBN 0-7923-1396-8
11. R.T. Haftka and Z. Gürdal: *Elements of Structural Optimization.* 3rd rev. and exp. ed. 1992 ISBN 0-7923-1504-9; Pb 0-7923-1505-7
12. J.R. Barber: *Elasticity.* 1992 ISBN 0-7923-1609-6; Pb 0-7923-1610-X
13. H.S. Tzou and G.L. Anderson (eds.): *Intelligent Structural Systems.* 1992 ISBN 0-7923-1920-6
14. E.E. Gdoutos: *Fracture Mechanics.* An Introduction. 1993 ISBN 0-7923-1932-X
15. J.P. Ward: *Solid Mechanics.* An Introduction. 1992 ISBN 0-7923-1949-4
16. M. Farshad: *Design and Analysis of Shell Structures.* 1992 ISBN 0-7923-1950-8
17. H.S. Tzou and T. Fukuda (eds.): *Precision Sensors, Actuators and Systems.* 1992 ISBN 0-7923-2015-8
18. J.R. Vinson: *The Behavior of Shells Composed of Isotropic and Composite Materials.* 1993 ISBN 0-7923-2113-8

Kluwer Academic Publishers – Dordrecht / Boston / London

Mechanics

***SOLID* MECHANICS AND ITS APPLICATIONS**

Series Editor: G.M.L. Gladwell

19. H.S. Tzou: *Piezoelectric Shells.* Distributed Sensing and Control of Continua. 1993 ISBN 0-7923-2186-3
20. W. Schiehlen: *Advanced Multibody System Dynamics.* Simulation and Software Tools. 1993 ISBN 0-7923-2192-8
21. C.-W. Lee: *Vibration Analysis of Rotors.* 1993 ISBN 0-7923-2300-9

Kluwer Academic Publishers – Dordrecht / Boston / London

GPSR Compliance

The European Union's (EU) General Product Safety Regulation (GPSR) is a set of rules that requires consumer products to be safe and our obligations to ensure this.

If you have any concerns about our products, you can contact us on ProductSafety@springernature.com

In case Publisher is established outside the EU, the EU authorized representative is:

Springer Nature Customer Service Center GmbH
Europaplatz 3
69115 Heidelberg, Germany

Batch number: 10607320

Printed by Printforce, the Netherlands